BREAKNECK

BREAKNECK

CHINA'S QUEST TO ENGINEER THE FUTURE

★ ★ ★ ★

DAN WANG

ALLEN LANE
an imprint of
PENGUIN BOOKS

ALLEN LANE

UK | USA | Canada | Ireland | Australia
India | New Zealand | South Africa

Allen Lane is part of the Penguin Random House group of companies
whose addresses can be found at global.penguinrandomhouse.com.

Penguin Random House UK
One Embassy Gardens, 8 Viaduct Gardens, London SW11 7BW

penguin.co.uk

Penguin
Random House
UK

First published in the United States of America by W. W. Norton & Company, Inc. 2025
First published in Great Britain by Allen Lane 2025
009

Printed and bound in Great Britain by Clays Ltd, Elcograf S.p.A.

The authorized representative in the EEA is Penguin Random House Ireland,
Morrison Chambers, 32 Nassau Street, Dublin D02 YH68

A CIP catalogue record for this book is available from the British Library

ISBN: 978-0-241-72917-5

Penguin Random House is committed to a sustainable future
for our business, our readers and our planet. This book is made from
Forest Stewardship Council® certified paper.

MIX
Paper | Supporting
responsible forestry
FSC
www.fsc.org   FSC® C018179

To my parents

CONTENTS

Introduction   ix

Chapter 1:   **Engineers vs. Lawyers**   1

Chapter 2:   **Building Big**   21

Chapter 3:   **Tech Power**   57

Chapter 4:   **One Child**   95

Chapter 5:   **Zero-Covid**   129

Chapter 6:   **Fortress China**   171

Chapter 7:   **Learning to Love Engineers**   209

             Acknowledgments   235

             Notes   239

             Suggestions for Further Reading   259

INTRODUCTION

E ACH TIME I SEE a headline announcing that officials from the United States and China are once more butting heads, I feel that the state of affairs is more than just tragic; it is comical, too, because I am sure that no two peoples are more alike than Americans and Chinese.

A strain of materialism, often crass, runs through both countries, sometimes producing veneration of successful entrepreneurs, sometimes creating displays of extraordinary tastelessness, overall contributing to a spirit of vigorous competition. Chinese and Americans are pragmatic: They have a get-it-done attitude that occasionally produces hurried work. Both countries are full of hustlers peddling shortcuts, especially to health and to wealth. Their peoples have an appreciation for the technological sublime: the awe of grand projects pushing physical limits. American and Chinese elites are often uneasy with the political views of the broader populace. But masses and elites are united in the faith that theirs is a uniquely powerful nation that ought to throw its weight around if smaller countries don't get in line.

I came to this view as a Canadian who has spent almost equal amounts of time living in the United States and China. To me, these two countries are thrilling, maddening, and, most of all, deeply bizarre. Canada is tidy. I sometimes find myself relaxing as soon as I cross into its borders. Drive around America and China, on the other hand, and you'll see people and places that are utterly deranged. That's not a reproach. These two countries are messy in part because they are both engines for global change. Europeans have a sense of optimism only about the past, stuck in their mausoleum economy because they are too sniffy to embrace American or Chinese practices. And the rest of the world is either too mature or too young to match the impact of these two superpowers. It is Americans and Chinese—Silicon Valley, Shenzhen, Wall Street, and Beijing—that will determine what people everywhere will think and what they will buy.

They are not the only two countries in the world that matter. Far from it. But if we don't understand how the United States and China function and interact, then in large part we won't quite understand many of the biggest changes in the world. The two countries are reconfiguring the international order and each other too. Seeing China more clearly—its dazzling strengths, appalling weaknesses, and everything in between—also helps us to see America more clearly.

To understand China, we must start in the country's most riveting city: Beijing.

Beijing enthralls not because it is nice but because it isn't. By most measures, life in Beijing is dreary. It is in China's arid north, where dust storms descend every so often upon the city's twisting alley homes, dating from imperial times, or gray apartment blocks, built in the Soviet style. In the last decade or so, the state has bricked up many of its liveliest sites, including its many bars and roadside barbecues, turning the city into a no-fun zone. Want to take your life into your hands? Try braving the cars that speed through Beijing's gigantic roads. Much like Moscow or Pyongyang, its avenues feel like they were built for army parades rather than for normal life. Really, everything that can go wrong in urban design has gone wrong in Beijing.

But the capital is also a city of gravity and substance. Beijing attracts many of China's smartest people, including scientists, technology leaders, and those seeking to advance in the Communist Party. The po-faced members of the Politburo don't fool around. Greatness isn't only a slogan for them: It's a full-on, life-or-death pursuit. Beijing, for the rest of this book, stands in as the Communist Party and the central government. China's leaders are driven by intense paranoia, doing everything they can to control the future.

My parents and I emigrated from China to Canada when I was seven. During high school, we moved to the woodsy suburbs of Philadelphia (where my mom and dad still live). After going to New York for college and Silicon Valley for work, I returned to China to investigate its technology developments. I learned to appreciate something vital: The country is always in motion. Living in Hong Kong, Beijing, and then Shanghai was a good education not only because these were China's most prosperous economic zones. For six years, I lived through a period of economic dynamism that gave way to smothering political repressiveness. I experienced top leader Xi Jinping's ongoing mobilization of the country for great-power competition. I tracked the expanding web of US restrictions on Chinese tech companies, as well as their struggle to escape from American restraints. And I endured all three years of Xi's pursuit of zero-Covid, which started impressively until it plunged the country into broad misery.

The Chinese state builds gleaming public works and doesn't flinch from locking up ethnic minorities or locking down whole cities. Too many outsiders see only the enrichment or the repression. Living there puts you face to face with both a sustained rise in living standards and the authoritarian pulses emanating out of Beijing. It became no contradiction for me to appreciate that things are getting better *and* getting worse. I saw how China is made up of both strong entrepreneurs *and* a strong government, with a state that both moves fast and breaks things *and* moves fast and breaks people.

I was the technology analyst at Gavekal Dragonomics, an investment research firm serving a financial audience. We were a small team of analysts managed by editors who used to be economics journalists. My task consisted of writing research notes for hedge funds, endowments, and other asset managers hungry for China analysis. Dragonomics research wasn't focused on particular companies but rather on more ambitious macro questions about the direction China was heading and what it means for the world. Portfolio managers aren't shy about getting to the heart of the matter, asking me: Can China's political system *really* breed tech giants? Will advanced manufacturing succeed when the rest of the world is throwing up trade barriers? How does a faltering economy affect Beijing's designs on Taiwan?

If I didn't offer good answers, the conversations could feel like a Socratic beating rather than a collegial chat. Though hedge fund managers can be obnoxious, I found conversing with them to be valuable. Folks in finance easily turn philosophical, pushing me to sharpen my views on important questions. I worked hard to decipher where Xi was taking China, which meant reading party texts, no matter how arcane, and visiting different regions, no matter how obscure.

By traveling as often as possible to smaller cities—some that are little more than urbanized industrial parks—I grasped something that most Americans, and even many Chinese, do not: Going to little-known cities in China is *fun*. Wherever I went I found amazing food, bizarre sights, and memorable people. I saw that China had greater dynamism than acknowledged by most headlines about the country, which fixate on Beijing's political machinations. Just imagine what the rest of the world would miss if they understood the United States exclusively through developments from Washington, DC.

Everywhere I felt China's breathless and, at times, reckless speed. I tried to capture the country's shifts and tussles, buffeted by a pandemic and a darkening international environment, by writing an annual letter. These were a journal of sorts to record everything I observed and felt. In 2020, I wrote about reading every Xi Jinping speech in *Seeking Truth*, the Communist Party's flagship theory magazine; in 2021, the differences between Hong Kong, Beijing, and Shanghai; and in 2022, what it was like to wander through the mountains of Yunnan province—whose north is historic Tibet and whose south feels like Thailand—during the worst period of zero-Covid.

I thought constantly about the United States. It wasn't only that the Trump administration was prosecuting a trade and technology war; Beijing holds America steadfastly in its gaze. China's leaders are ready to learn from Europe, Japan, Singapore, and many others, sure. But they have looked up to the United States more than any other country, benchmarking themselves against the world's preeminent power.

It is almost uncanny how much the United States and China have been complementary of each other. It was no accident that the two countries established, for a few decades, an economic partnership that worked tremendously well for American consumers and Chinese workers. But on a political level, these two systems are a study in contrasts. While the United States reflects the virtues of pluralism and protection of individuals, China revealed the advantages and perils that come from moving quickly to achieve rapid physical improvements.

Over the past four decades, China has grown richer, more technologically capable, and more diplomatically assertive abroad. China learned so well from the United States that it started to beat America at its own game: capitalism, industry, and harnessing its people's restless ambitions. If you want to appreciate what Detroit felt like at its peak, it's probably better to experience that in Shenzhen than anywhere in the United States.

As China emulated America's past successes, the US government got busy undermining its own strengths. A procedure-obsessed left conspired with a thoughtlessly destructive right to constrain the government. Neither the left nor the right allows the state to deliver essential goods expected by the public. The Biden administration may have ushered through historic bills on industrial policy, but executive agencies were so obsessed with procedural concerns that little building actually took place before voters reelected Donald Trump, who has threatened to cancel many of these projects. The United States is still a superpower that is able to outclass China on many dimensions. But it is also in the grips of an ineffectual state where people are increasingly concerned with safeguarding a comfortable way of life.

Americans used to love the great opportunity that China represented. Nearly a century ago, they were wartime allies, with ties cemented by cultural connections and business relationships. Today, natural amity is being crowded out by mutual mistrust. Beijing and Washington are competing with each other economically, technologically, and diplomatically, casting a pall on those of us connected to both countries. In 2022, Beijing's censors blocked the personal website where I publish my annual letters. The Great Firewall tends to block access to big platforms like the *New York Times,* not little sites like mine. That week, I had to seek out the Canadian consul general to ask whether I needed to organize my departure from China. Beijing had already detained two Canadians in response to Canada's arrest of a prominent Chinese businesswoman. Many Americans who previously traveled to China for business and pleasure have lost their enthusiasm for visits.

We are now in an era where the two countries regard each other with suspicion, and often animosity. Like China, the United States is able to move fast and break people, dealing tremendous brutality at home and abroad when it feels threatened. A paramount question of our times is whether hostility between China and the United States can stay at a manageable simmer. Because if it boils over, they will devastate not only each other but also the world.

The best hedge I know against heightening tensions between the two superpowers is mutual curiosity. The more informed Americans are about Chinese, and vice versa, the more likely we are to stay out of trouble. The starkest contrast between the two countries is the competition that will define the twenty-first century: an American elite, made up of mostly lawyers, excelling at obstruction, versus a Chinese technocratic class, made up of mostly engineers, that excels at construction. That's the big idea behind this book. It's time for a new lens to understand the two superpowers: China is an *engineering state*, building big at breakneck speed, in contrast to the United States' *lawyerly society*, blocking everything it can, good and bad.

Breakneck is the story of the Chinese state that yanked its people into modernity—an action rightfully envied by much of the world—using means that ran roughshod over many—an approach rightfully disdained by much of the world. It is also a reminder that the United States once knew the virtues of speed and ambitious construction. Traversing dazzling metropolises and gigantic factories, *Breakneck* will illuminate the astounding progress and the dark underbelly of the engineering state. The lawyerly society has virtues, too, to teach China. Each superpower offers a vision of how the other can be better, if only their leaders and peoples care to take more than a fleeting glance.

BREAKNECK

CHAPTER 1

ENGINEERS VS. LAWYERS

SILICON VALLEY CAN BE an amazingly drab place. The peninsula south of San Francisco has natural beauty, with rolling hills and coastal views, but you strain to see them beyond so many corporate parking lots. Mountain View and Menlo Park are bizarrely full of rug shops, so when I walk through the towns that host the headquarters of AI leaders and some of the richest companies in the world, I often find myself wondering, "*This* is the beating heart of our technologically accelerating civilization?"

Each time I flew from California to Hong Kong or Shanghai, I felt almost unnerved to encounter functional infrastructure. Going from the airport into a subway (rather than an Uber) is an outstanding way to be welcomed to Asia. I would take a moment to savor a clean station, brightly lit, with trains running every few minutes, which would drop me off at a downtown filled with vibrant commercial areas—another feature that San Francisco lacks. Life in the Bay Area, an economic dynamo in America's richest state, can feel awfully dysfunctional. San Francisco has been unable to serve its homeless population, and even many wealthy people have to keep a generator for their extraordinarily expensive houses because the state can't keep the lights on.

The contradiction of the Bay Area, this red-hot center of corporate value creation that is surrounded by dysfunction, fuels the inquiry of this book. When I departed from Silicon Valley for China in 2017, it felt clear that the United States had lost something special over the past four decades. While China was building the future, America had become physically static, its innovations mostly bound up in the virtual and financial worlds.

Looking at these two countries, I came to realize the inadequacy of twentieth-century labels like capitalist, socialist, or, worst of all, neoliberal. They are no longer up to the task of helping us understand the world, if they ever were. Capitalist America intrudes upon the free market with a dense program of regulation and taxation while providing substantial (albeit imperfect) redistributive policies. Socialist China detains union organizers, levies light taxes, and provides a threadbare social safety net. The greatest trick that the Communist Party ever pulled off is masquerading as leftist. While Xi Jinping and the rest of the Politburo mouth Marxist pieties, the state is enacting a right-wing agenda that Western conservatives would salivate over: administering limited welfare, erecting enormous barriers to immigration, and enforcing traditional gender roles—where men have to be macho and women have to bear their children.

China is an *engineering state*, which can't stop itself from building, facing off against America's *lawyerly society*, which blocks everything it can.

Engineers have quite literally ruled modern China. As a corrective to the mayhem of the Mao years, Deng Xiaoping promoted engineers to the top ranks of China's government throughout the 1980s and 1990s. By 2002, *all* nine members of the Politburo's standing committee—the apex of the Communist Party—had trained as engineers. General Secretary Hu Jintao studied hydraulic engineering and spent a decade building dams. His eight other colleagues could have run a Soviet heavy-industry conglomerate: with majors in electron-tube engineering and thermal engineering, from schools like the Beijing Steel and Iron Institute and the Harbin Institute of Technology, and work experience at the First Machine-Building Ministry and the Shanghai Artificial Board Machinery Factory.

Xi Jinping studied chemical engineering at Tsinghua, China's top science university. For his third term as the Communist Party's general secretary starting in 2022, Xi filled the Politburo with executives from the country's aerospace and weapons ministries. In the United States, it would be as if the CEO of Boeing became the governor of Alaska, the chief of Lockheed Martin became the secretary of energy, and the head of NASA was governor of a state as large as Georgia. China's ruling elites have practical experience managing megaprojects, suggesting that China is doubling down on engineers—and prioritizing defense—more than ever.

What do engineers like to do? Build. Since ancient times, the emperors have tried to tame the mighty rivers that sweep away not only farmland, but also imperial reigns. In modern times, new public works—roads, bridges, tunnels, dams, power plants, entire new cities—are the engineering state's solution to any number of quandaries. Since 1980, after Deng's reforms began, China has built an expanse of highways equal to twice the length of the US systems, a high-speed rail network twenty times more extensive than Japan's, and almost as much solar and wind power capacity as the rest of the world put together. It's not only the government that is fixated on production; the corporate sector is made up of overactive producers too. A rough rule of thumb is that China produces one-third to one-half of nearly any manufactured product, whether that is structural steel, container ships, solar photovoltaic panels, or anything else.

When Chinese point to new cities that shimmer at night with drone displays, or metropolises connected to each other by a glistening high-speed rail network, their pride is real. Call it propaganda of the deed, but one way to impress a billion-plus people is to pour a lot of concrete.

The United States, by contrast, has a government of the lawyers, by the lawyers, and for the lawyers. Five out of the last ten presidents attended law school. In any given year, at least half the US Congress has law degrees, while at best a handful of members have studied science or engineering. From 1984 to 2020, every single Democratic presidential and vice-presidential nominee went to law school, but they make up many Republican Party elites as well as the top ranks of the civil service too. By contrast, only two American presidents worked as engineers: Herbert Hoover, who built a fortune in mining, and Jimmy Carter, who served as an engineering officer on a nuclear submarine. Hoover and Carter are remembered for many things, especially for their dismal political instincts that produced thumping electoral defeats.

Lawyers have so many tools available to delay or prevent building. You don't just feel the difference going from the lawyerly society to the engineering state: You saunter, tread, and amble upon its works. Americans no longer manufacture well or build public works on reasonable timelines. US infrastructure is falling into a pitiable state while China is building new systems of subways, bridges, and highways. Over the past three decades, while Chinese manufacturers have been going from strength to strength . . . well, let's just say that American automakers and chipmakers haven't exactly covered themselves in glory. China's political system is geared toward delivering monumental projects, such that the slightest economic tremble is enough to push Beijing to announce a mammoth plan for new public works. That's one reason that the phrase "housing crisis" has evoked, over the past several years, a collapse of home prices for Chinese and spiraling unaffordability for Americans.

Lawyers enable some of the success of Silicon Valley. You can't build companies worth trillions without legal protections. But lawyers are also part of the reason that the Bay Area and much of the country are starved of housing and mass transit. The United States used to be, like China, an engineering state. But in the 1960s, the priorities of elite lawyers took a sharp turn. As Americans grew alarmed by the unpleasant by-products of growth—environmental destruction, excessive highway construction, corporate interests above public interests—the focus of lawyers turned to litigation and regulation. The mission became to stop as many things as possible.

As the United States lost its enthusiasm for engineers, China embraced engineering in all its dimensions. Its leaders aren't only civil or electrical engineers. They are, fundamentally, social engineers. Emperors didn't hesitate to entirely restructure a person's relationship to the land, ordering mass migration into newly opened territories and conscripting the people to build great walls or grand canals. Modern rulers are here, too, far more ambitious than the emperors of the past. The Soviet Union inspired many of Beijing's leaders with a love of heavy industry and an enthusiasm to become engineers of the soul—a phrase from Joseph Stalin repeated by Xi Jinping—heaving China's population into modernity and then some.

Modern China has many tools of social control. Within living memory, most Chinese residents worked inside a *danwei,* or work unit, which governed one's access to essentials like rice, meat, cooking oil, and a bicycle. Many people still live under the strictures of the *hukou,* or household registration, an aim of which is to prevent rural folks from establishing themselves in cities by restricting education and health care benefits to their hometown. Controls are far worse for ethnoreligious minorities: Tibetans are totally prohibited from worshipping the Dalai Lama, and perhaps over a million Uighurs have spent time in detention camps that attempt to inculcate Chinese values into their Muslim faith.

The engineering state can be awfully literal minded. Sometimes, it feels like China's leadership is made up entirely of hydraulic engineers, who view the economy and society as liquid flows, as if all human activity—from mass production to reproduction—can be directed, restricted, increased, or blocked with the same ease as turning a series of valves.

Can a government be too efficient? Six years in China taught me that the answer is yes, when it is unbounded by citizen input. There are many self-limiting aspects of a system that makes snap decisions with so little regard for people. This book reveals good things that the engineering state does: running functional cities, building up its manufacturing base, and spreading material benefits pretty widely throughout society. But I also lived through things that no other state would have attempted, like holding on to a zero-Covid strategy until it drove the country mad. The fundamental tenet of the engineering state is to look at people as aggregates, not individuals. The Communist Party envisions itself as a grand master, coordinating unified actions across state and society, able to launch strategic maneuvers beyond the comprehension of its citizens. Its philosophy is to maximize the discretion of the state and minimize the rights of individuals.

Engineers often treat social issues as math exercises. Does the country have too many people? Beijing's solution was to prohibit families from birthing more than one child—the subject of my fourth chapter—through mass sterilization and abortion campaigns, as the central government ordered in 1980. Is the novel coronavirus spreading too quickly? Build new hospitals at breathtaking speed, yes, but also confine people to their homes, as Wuhan, Xi'an, and Shanghai did to millions of people over weeks, which I cover in the fifth chapter. There is no confusion about the purpose of zero-Covid or the one-child policy: The number is right there in the name.

China's economy isn't immune to engineering either. When

Beijing grew uncomfortable with the debt levels of real estate developers in 2021, the state forced so many of them into distress that it triggered a prolonged slump in homebuyer confidence. Around the same time, Xi hurled a series of regulatory thunderbolts at China's high-flying tech companies, including Didi, the country's largest ride-hailing company, and Ant Financial, the payments company owned by Jack Ma, China's best-known entrepreneur. Chinese tech founders (and their investors) were astonished to discover that Xi Jinping could erase a trillion dollars from corporate valuations over the course of just a few months. The leadership thought it was straightforward to reorient the nation's tech priorities away from consumer platforms and toward science-based industries, like semiconductors and aviation, that serve the nation's strategic needs. Beijing took years to appreciate how its actions had scared the daylights out of entrepreneurs and investors.

When you travel around China, it's staggering to see how much the engineering approach has accomplished over the past four decades. Then there's the part you can't see. As impressive as China's railways and bridges may be, they carry enormous levels of debt that drag down broader growth. Manufacturers produce so many goods that China's trade partners are now grumbling for protection. The social-engineering experiment known as the one-child policy has accelerated the country's demographic decline. And China's economy would be in better shape if Beijing hadn't triggered an implosion of its property sector, smothered many of its most dynamic companies, and persisted in trying to push out the coronavirus.

Well-to-do people professionals who thought themselves secure in their jobs in finance or consumer internet faced a rude shock when Xi's displeasure with these sectors caused rippling job losses. No US president has so much ability to overturn the lives of the rich. By contrast, in China, many pillars of society are liable to blow over when winds from Beijing shift direction, contributing to a sense of precarity among even the country's elites. Since China doesn't have many legal protections, not even its rich are well protected.

Engineers go hard in one direction, and if they perceive something isn't working, they switch with no loss of speed toward another. They don't suffer criticism from humanist softies. Change in China can be so dramatic because so few voices are part of the political process. To a first approximation, the twenty-four men who make up the Political Bureau (the highest echelon of the Communist Party, usually shortened to Politburo) are the only people permitted to do politics. Once they've settled questions of strategy, the only remaining task is for the bureaucracy to sort out the details. But when it makes mistakes, it can drag nearly the entire population into crisis.

To capture both the traumatic aspects of the engineering state and its capacity to produce great pride, I like to think of a hypothetical question: What was the worst year to be born in modern China?

A strong contender, I believe, is 1949, the year Mao Zedong founded the People's Republic. A person born that year—let's call her Lu—would live through several of China's utopian experiments, which curdled into terror campaigns led by the state. Lu would be born into a country torn up by Japan's invasion and a civil war, but hopeful about Mao's promise of communism. Around age ten, Lu would suffer some degree of food shortage as she lived through Mao's scheme to get industrialized quick. That was the Great Leap Forward, when tens of millions perished from agricultural collectivization, quack agronomy, natural disasters, and Mao's order to melt down household tools for the metal, all leading to the sort of mass starvation that forced people to forage tree bark to survive. At age eighteen, Lu might have just missed her chance to attend college as Mao shut down higher education. "Rebellion is justified," he told students while launching the Cultural Revolution. "Bombard the headquarters," he instructed youths while sending them into the countryside.

If Lu decided to have a child after the age of thirty, she would have run into the one-child policy. Over the policy's three-and-a-half-decade duration, China conducted nearly as many abortions, according to official figures, as the present population of the United States. If Lu had given birth at age twenty, her child might have attended college in 1989. That spring and summer, students led protests throughout the country, most prominently in Beijing. By June, Deng Xiaoping declared martial law and deployed the army to mow down students from the country's most elite colleges. A few years after the killings around Tiananmen Square, China's economic boom began in earnest. But as Lu turned seventy and entered the twilight of her life, she would feel one last spasm of a state-led terror campaign: lockdowns in the pursuit of zero-Covid. Depending on whether Lu lived in an unlucky city, she might not have been able to leave her residence for weeks.

But change the year of birth by a decade and outcomes can shift spectacularly.

Someone born in 1959 would have no memory of famine. Call this luckier citizen Yao. By the time he turned eighteen, Mao would have died, and Yao could have earned a spot in university just as Deng was reopening the schools. As he turned forty and entered the prime of his career, he might have established a business that capitalized on China's entry into the World Trade Organization. Also around then, if he were an urban resident, Yao would catch China's housing privatization. As the state moved to dismantle socialism, it offered homes to urban workers for a song. It was one of the greatest wealth transfers in history: If Yao was among the elites who owned real estate in Beijing and Shanghai, which grew into two of the world's most expensive cities, he could have become prodigiously wealthy.

Not everyone born in 1949 suffered terribly and not everyone born in 1959 lived comfortably. But the engineering state is characterized by peculiarly jerky rhythms, in which the decade of birth might determine whether a person stumbles into great wealth or a mass grave.

The generation of Chinese born in the 2000s are somewhere in between these extremes. College graduates have, in recent years, contended with record high youth unemployment while their parents mourn falling property values. For one group of online nationalists, nicknamed "little pinks," China can't stop winning. The collapse of the property sector was good and necessary, they contend, because investment is going into manufacturing. And if China's broader economy is weak, they say that China's economic woes are caused by the United States.

The second argument is just ridiculous. Yes, tariffs and technology controls have hurt Chinese firms. But what is the US government's damage to China's economy next to the Politburo's shock tactics? That the United States is able to hobble China's growth is believable only inside the country's highly censored information environment.

But the little pinks have a line of argument that tickles me. "Look at those Americans," a few say, "who have no high-speed rail or gleaming skyscrapers like we do. Their only skill is blocking themselves, which they are now doing to us." Little pinks are wrong to say that the United States is powerful enough to tank China's economy; they're not wrong that the United States blocks itself.

・　・　・

THE YEAR 2008 OFFERS a direct comparison between California's speed and China's speed. That year, California voters approved a state proposition to fund a high-speed rail link between San Francisco and Los Angeles; also that year, China began construction of its high-speed rail line between Beijing and Shanghai. Both lines would be around eight hundred miles long upon completion.

China opened the Beijing–Shanghai line in 2011 at a cost of $36 billion. In its first decade of operation, it completed 1.35 billion passenger trips. California has built, seventeen years after the ballot proposition, a small stretch of rail to connect two cities in the Central Valley, neither of which are close to San Francisco or Los Angeles.

The latest estimate for California's rail line is $128 billion. Why does it cost so much? Partly because some politicians have demanded that the train add a stop in their district, forcing the line to take a more tortuous route through an extra mountain range. And partly because California's rail authority prefers to tout the number of high-paying jobs it is creating rather than the amount of track it has been laying. The first segment of California's train will start operating, according to official estimates, between 2030 and 2033. Which means that the *margin of error* for estimating when a partial leg of California's high-speed rail will open is the same as the time it took China to build the entire Beijing–Shanghai line.

The United States wasn't always like this. American mayors and governors used to love attending ribbon-cutting ceremonies. These are now few and far between. American cities have broadly failed to build adequate housing or infrastructure. What they do complete—a public bathroom, a bus stop, or, my goodness, a subway station—arrives embarrassingly late or over budget. Americans live today in the ruins of an industrial civilization, whose infrastructure is just barely maintained and rarely expanded.

Once upon a time, America, too, had the musculature of an engineering state, building mighty works throughout the country: lengthy train tracks, gorgeous bridges, beautiful cities, weapons of war with terrible power, and missions to the moon. George Washington was a general, the first of many national security types who appreciated the value of building. As a young army officer, Dwight Eisenhower spent two months driving, or, more precisely, juddering, from coast to coast on unpaved roads. As president, he built the Interstate Highway System. When the United States had surging population and economic growth through the nineteenth century, political elites agreed that its vast territories needed canals, rails, and highways. Some of the leading figures in the Progressive Era embraced social engineering—and they conducted enough eugenics experiments to prove it.

China today resembles the United States of a century ago while it was proving itself to be a superpower. But America's construction boom slowed down after the 1960s. What happened next? The lawyers.

In the 1960s, parts of the United States had grown into a frightful place. Oil platforms discharged petroleum into the sea, a foul smog settled over cities, and factories leaked so many chemicals that even rivers combusted. Urban planners rammed highways through urban neighborhoods. Legal discrimination segregated people by race and blocked them from exercising the right to vote. The public soured on the idea of broad deference to US technocrats and engineers: urban planners (who were uprooting whole neighborhoods), defense officials (who were prosecuting the war in Vietnam), and industry regulators (who were cozying up to companies).

Students at elite law schools, especially Yale and Harvard, sprang up to act. Students founded environmental organizations around the rallying cry of "Sue the bastards!" (referring to government agencies). Through the 1970s, both the American left and the right worked harmoniously to constrain government effectiveness. Liberal activists like Ralph Nader declared themselves to be watchdogs of government, constantly filing lawsuits. Ronald Reagan returned the compliment when he replied, "Government is the problem, not the solution." The lawyerly society grew out of a necessary corrective to the United States' problems of the 1960s. Unfortunately, it has become the cause of many of its present problems.

As a fellow at Yale Law School's Paul Tsai China Center, I peered at the lawyerly society from within one of its high temples. The law students I got to know are smart, friendly, and most of all, ambitious. They are good at climbing prestige ladders—joining a law review board as a student and clerking for a federal judge after graduation. Yale Law students mostly lean left, but there are also many conservatives among them. Case in point: J. D. Vance. Though the political views of law students may twist in unexpected directions, we should keep in view that they are entwined most firmly around a pillar of personal ambition.

More than any other group in the United States, lawyers are afforded license to be generalists, permitted to stomp into whichever intellectual realm pleases them. "American aristocracy," wrote Alexis de Tocqueville, "is not composed of the rich . . . but occupies the judicial bench and the bar." Lawyers have become even more powerful since Tocqueville wrote those words in 1833. In recent decades, lawyers have been able to muscle out economists even in economic policymaking. The Biden administration was staffed by many graduates of Yale Law, who were willing to ignore the logic of the invisible hand. Instead, they roll up their sleeves to perform surgery on the American economy, one case at a time, devising a subsidy scheme for one corporation or bringing an antitrust case against another. Lawyers create so many complications that the rules governing everything from health care and housing to banking have become incomprehensible.

The American courtroom is a battlefront to resolve political questions, in which judges are enlisted to rule on questions that most other countries leave to voters or regulators. When a political cause can't be won through the electoral process, lawyers sometimes seek a victory through the courts. Since the middle of the twentieth century, the American left pursued a "democracy by lawsuit" strategy that conservatives have revealed themselves to be no less capable at playing.

There are reasons to be happy for lawyers to have an outsized presence in American society. They are reliable conversationalists at cocktail parties, for example—much better than engineers or economists. More seriously, they help to maintain America's civic-mindedness and its commitment to laws. Many of them do important work: facilitating people's access to bankruptcy, divorce, or immigration services; helping secure civil rights; and working to protect wildlife and clean water. When the White House acts out of line, we hope that the judiciary will restrain it.

Though the lawyerly society corrected the problems of the past, it has produced two pathologies that weaken the United States today.

The first is an elevation of process over outcomes. In American government and society, designing new rules and committees have so often become the substitute for thinking hard about strategy and ends.

While engineers envision bridges, lawyers envision procedures. In a seminal paper titled "The Procedure Fetish," University of Michigan law professor Nicholas Bagley outlines how the federal government requires an agency to "conduct every conceivable study, ventilate every option, engage every identifiable stakeholder, and weather the most stringent judicial review before any of its actions, however trivial, could take effect." In the lawyerly society, a more rigorous process is the solution to any number of quandaries. To deal with a new problem, it designs another procedure, which usually entails longer bureaucratic deliberation, greater public discussion, and more intensive judicial review.

Lawyers have much more scope with the law to stop something rather than create something. Before a government agency can build anything—from simple things like a bike lane to more complex projects like California's high-speed rail—it ties itself down with mountains of procedure. The agency has to check so many boxes because it knows that a lawsuit could derail that bike lane if people are able to convince a judge it didn't study environmental problems hard enough. After exhaustive research and review, it is no wonder that little ends up built. Americans are left with decaying infrastructure, little new construction, and a deep sense that nothing is working.

It's not just the government. America's problem is the lawyerly *society*. The United States is unusual among Western countries for having so many lawyers: four hundred lawyers per hundred thousand people, which is three times higher than the average in European countries. Since lawyers are everywhere, proceduralism has reached everywhere, including universities and corporations. Anyone working in these today has seen how procedures become an end in themselves, such that people grow obsessed with their logic and forget about the outcome. Because who can keep the goal straight after the seventh monthly committee meeting?

The other problem of the lawyerly society is a systematic bias toward the well-off. Lawyers are too often servants of the rich. They help wealthy homeowners block construction projects or get creative with their taxes. It is sometimes puzzling to follow along intellectual property cases, many of which seem to be a thrilling game invented for lawyers. American judges have to deal with bewildering disputes, like hedge funds pursuing sovereign governments on debt payments. Litigation offers endlessly tantalizing possibilities for settling scores. And motivated parties are willing to pay top dollar for superstar lawyers. Lawyers aren't just defenders of the rich; many of them *are* the rich. "On Wall Street, Lawyers Make More Than Bankers Now" was a headline from the *Wall Street Journal* in 2023. "Pay for Lawyers Is So High People Are Comparing It to the NBA" claimed the *New York Times* in 2024.

America's dysfunctions are not obstacles for the rich. Though New York City has barely been able to extend its system of mass transit, real estate developers have been able to build skinny high-rises for the wealthy. Though California can't tame wildfires, the rich might be able to afford their own private firefighting services. The poor—those buried under paperwork trying to apply for SNAP benefits, who have to take dilapidated public transit and who would most benefit from new construction—are the ones who suffer most from the lawyerly society's failures.

I am not saying, as Shakespeare's Dick the Butcher snipes in *Henry VI, Part 2*, that "the first thing we do, let's kill all the lawyers." The system of checks and balances has been, and is, fundamental to the success of the United States. Since the government is capa-

ble of wielding terrible power, judges and the law are often the last
and best hope against abuses. But the United States will not remain
a great power if it caters primarily to the wealthy. Its failure to build
enough has hurt working people and makes the country feel like a
low-agency society.

The engineering state is more than autocracy or technocratic high
modernism. China has succeeded better than any other authoritarian country in history at combining economic growth with political
control. The Communist Party has relentlessly broken up entrenched
interests, partly to prevent rich people from gaining political power
and partly to spread material benefits throughout the country. Its
rise suggests that a country can grow powerful when it trains a lot of
engineers and puts them to work, even under less-than-great institutional arrangements. In the words of one 1991 paper written by a
trio of economists, "Our evidence shows that countries with a higher
proportion of engineering college majors grow faster; whereas countries with a higher proportion of law concentrators grow more slowly."
Engineers are part of the reason that China has grown so much
wealthier, despite its wavering commitment to secure property rights.
An engineer mindset is also part of the reason that the state skirted so
close to apocalypse—in the case of the Cultural Revolution—before it
achieved a growth miracle.

The lawyerly society doesn't have such dramatic shifts. It is made
up of democracy, pluralism, vetocracy, and not only these things. The
lawyerly society also includes a commitment to proceduralism and
protecting wealth. Economically, the United States has experienced
strong economic growth relative to other Western countries combined with astonishingly successful corporate value creation. But in
political terms, this obsession with process over outcomes has made
Americans lose faith that the government can meaningfully improve
their lives. I want the US government to earn back that faith. To do
so, it will need to recover some of its engineering prowess and make

room for nonlawyers among its ruling elites. It will require the United States to build again, creating a momentum and the sense of optimism for the future that many Chinese have felt over the past two decades.

• • •

THE REASON WE HAVE to get smarter about both the United States and China is not because they are fascinating intellectual puzzles in themselves. It is because the two superpowers are uneasily circling each other, reorienting their economies and national security apparatuses to prepare for conflict.

As China and the United States gear up for competition and conflict, we need fresh ways—using terms that are not amalgamations from political science texts—to think about how both countries function and how they fail. The engineering state and the lawyerly society are not the only ways to understand the two countries; old labels like "autocratic" or "capitalist" still have some use, of course. I want to be inventive and even playful with these terms in the interest of encouraging mutual curiosity between the two countries.

The United States has immense advantages over China: robust economic growth, an expanding and more youthful population, innovation in digital technologies, a larger network of alliances, and more. But we need to recognize that the engineering state has a giant advantage: China can build. That will matter if the two countries ever decide, in an apocalyptic scenario, to go to war. No military can be powered by artificial intelligence alone; it will need drones and munitions. And the engineering state is better set up to produce these in overwhelming quantity.

Over the past decade, the United States brought lawyers to a technology fight. The first Trump administration blacklisted scores of Chinese tech companies. In the Biden administration, the ranks of the National Security Council and the Department of Commerce were filled with graduates of elite law schools, including their department heads. Lawyers have designed exquisite webs of technology controls, ensnaring Chinese chipmakers, telecommunications firms, and any company hoping to deploy AI. Rather than halt Chinese technology leaders in their tracks, these legal controls have riled them up. When Xi started his third term in 2022, he didn't stack the Politburo with clever lawyers able to deliver a really good rebuttal. He filled it with scientists and engineers. They will help to design the Fifteenth Five-Year Plan, which will place even greater emphasis on building technological strength.

A contest between a literal-minded dragon and lawyerly weenies wouldn't be a fair fight. The struggle is more complex than that. Whether one can outlast the other will depend not only on physical dynamism or technological prowess. It will depend on governance—which country can do a better job managing its affairs over the next century.

As best as I can tell, the United States and China are both racing to erode their governance capabilities. Xi Jinping has forcefully centered the political decision-making process on himself, demonstrating that he intends to rule the Communist Party for as long as he pleases. The American government, meanwhile, has been mired in ineffectualness. For decades, the American right connived to drown the government in a bathtub while the left was strangling it with rules and lawsuits. The left has barely shown resolve to reform creaky institutions, and the second Trump administration behaves as if it must destroy the government in order to save it.

But there is hope for everyone. The most important thing that China and the United States share is a commitment to transformation. China is led by a Leninist party whose core aim is to mobilize society toward modernization. Its propaganda organs stage centralized campaigns of inspiration toward the centenary goal to achieve, by 2049, "a modern socialist country" and "the great rejuvenation of the Chinese nation." The US commitment is more open-ended, inherent to the experiment to keep democracy going. That has been partly deformed, but we should revive the dream that government of the people, by the people, and for the people shall not perish.

I am not impressed with California's governance. But I want to reveal there is one aspect of the California attitude that I fully imbibe: I am a sunny optimist for the future, with faith that both societies can change for the better. Both countries are in a state of becoming, which means that either of the superpowers are able to tilt away from their present, bad courses.

If Americans look deeply into China, they will find reflections of its lost powers. China, right now, is in the midst of pursuing its own Great Society, where even its poorest provinces have impressive levels of physical dynamism. Delivering the goods is part of why consent of the governed is still pretty strong in China. I saw that for myself when I spent five days furiously pedaling through the jagged mountains of Guizhou.

CHAPTER 2

BUILDING BIG

MY MOST VIVID ENCOUNTER with the engineering state occurred, in classic Chinese fashion, on a bicycle.

In the summer of 2021, I traveled with two friends deep into China's southwest. Over five days, we cycled nearly four hundred miles through Guizhou province and arrived in the city of Chongqing. Rather than riding a Flying Pigeon—the comfortable but single-geared bike from the Maoist era, available only in black—I was flying through on a Giant racing bike, which was fabulously strong and quick.

It was over this long ride that I started to realize how an examination of China's problems throws US problems into stark relief. Each time I left Beijing and Shanghai to enter more remote parts of the country, I was astonished by how even China's poorest provinces have better infrastructure than America's richest. The chief feature of the engineering state is building big public works, no matter the financial or human cost. For many people in Guizhou, it has produced an enthusiasm and an expectation for physical change, a feeling not often found among Americans today.

Mountains dominate Guizhou's landscape. They are made of karst stone, perplexed with intricacy. Even a decade ago, a cycling trip through Guizhou (pronounced *Gway-JOE*) might have been foolhardy. There just weren't enough adequate roads. It is China's fourth-poorest province and far away from prosperous coasts—a province, the saying goes, "where not three feet of land is flat, where not three days pass without rain, where not a family has three silver coins."

In the nineteenth century, one of the imperial cartographers sent by the Qing emperor to map the territory grew exasperated by his task. "Southern Guizhou has a multitude of mountain peaks jumbled together," he lamented. "They are vexingly numerous and ill-disciplined." Visitors did not always find locals to be hospitable. Much of Guizhou is settled by the Miao minority, which has historically resented the intrusion of China's Han-majority ethnic group.

Guizhou's insularity and mystery are the stuff of legend. One traveler in the ninth century wrote about an ordeal: While exploring the province, he chanced upon an elegant monastery. Ten nuns at once emerged, merrily inviting him inside their thatched cottages. They were excellent hosts, plying him with dried fruits. When he felt the scene to be too fantastic, the traveler braved the dismay of the nuns and abruptly departed. Once he returned to the boat, the crew confirmed his fears: The nuns were monkey tricksters who sometimes took on human form to entice people into their midst.

In the present century, the central government has lavished Guizhou with attention. Several of Guizhou's party chiefs have gone on to high positions in Beijing, including Hu Jintao, general secretary of the Communist Party before Xi Jinping. Chinese leaders are usually expected to administer a poor province before they can be promoted to the country's political pinnacle. In the United States, it would be as if politicians had to gain some experience in the Rust Belt or coal country before they could get anywhere near a cabinet position. Guizhou has received several big projects. The central government built the world's largest radio telescope—with an aperture measuring five hundred meters in diameter, named the Heaven's Eye—in a remote corner of the province. The state-owned distillery behind Maotai, the hundred-proof spirit made of sorghum, grew into one of China's most valuable companies. Its capital city of Guiyang now hosts several of the country's biggest data centers.

I went to Guiyang with my friends Christian Shepherd, a journalist from the United Kingdom then working for the *Financial Times*, and Teng Bao, who grew up in Florida and founded a tech company in Shanghai. A century ago, it would have taken weeks of travel along twisting roads to reach Guiyang from Shanghai. For my friends and me, it took a seven-hour ride on high-speed rail.

Guizhou was one of the last provinces to be connected to the national high-speed network. When the railway opened its first station in 2016, engineers had finally blasted tunnels through the mountains and erected enough sturdy bridges to span the gorges. On the train, Christian, Teng, and I reclined in comfortable seats, tucking our disassembled bikes in the back of the compartment, picking up snacks or water from the attendant's trolley when we wanted something. When we looked out the windows, the occasional blur of long tunnels hinted at the difficulty of the construction.

Christian is a great cyclist. Teng and I had more enthusiasm than experience. The three of us each packed a change of clothes, a first-aid kit, spare tires, and not much else. We stuffed our gear into sleek leather bags strapped on the back of our bikes. Then we were off. The plan was to reach our hostel accommodations by early evening each day, where we would wash our clothes in the sink, hang them out to dry, and then get up to do it all over again the next day.

Each day of cycling brought new thrills: spectacular landscapes, bridges and gorges that kept surpassing the last, waterfalls where we would occasionally linger. Our trek was tough—not because we were up against impassable roads or trickster monkeys, but because every day demanded the grind of pushing uphill. Guizhou's infrastructure was a cyclist's dream. On the first day of our trip, we cycled along a just-built highway not yet open to cars. That was our favorite moment: careening downhill at thrilling speeds amid luscious green mountains wreathed by bands of mist.

This bike ride was the greatest physical exertion of my life, as well as the most rewarding. We enjoyed not only the views but the food as well. Every few hours, we took a break along the side of the road. You expend an enormous amount of energy on a bike, so we would order bowls of noodles—spooning in the pungent pickles that make Guizhou cuisine so bracing—and then grab a vanilla ice cream bar before hopping back on our bikes. At night, we ordered local dishes: a fish stew full of sour pickles, braised goat, a salad of local herbs and roots, and rice balls (each the size of a lime) filled with sweet sesame, deep fried with savory pickles on the side.

If only that Qing cartographer could see Guizhou now. All sorts of new infrastructure are built into its countryside. On the third day, we came upon a sight nearly as strange as a monkey-filled phantasm. Teng was leading the three of us when he yelled, "Guitars!" When I raised my gaze, I saw that big guitar ornaments were hanging off of streetlamps. In the distance, I spied a hill topped by a giant rock guitar. It turned out that we were cycling through Zheng'an County, the self-styled guitar capital of the world. According to state media, one of every seven guitars made worldwide is produced in this township we passed through by chance.

That is another feature of the engineering state: Manufacturing hubs are everywhere, often making goods you don't expect.

Guizhou locals may be as surprised as anyone to host the world's guitar capital. Not many of them play the instrument. Zheng'an became a guitar hub because a lot of its residents had moved to coastal Guangdong for work, many of them finding employment by coincidence in guitar factories. Then the local government made a big effort to entice them to return to Guizhou as part of a policy to develop the interior. That effort coincided with a 2012 directive from the State Council (the executive agency of the central government) that encouraged manufacturers to relocate from coastal provinces to inland ones. The document had suggested that Guizhou pursue technologically intensive industries like aerospace or electric vehicle manufacturing. Instead, what Guizhou built was more suitable to its less-skilled realities: the Guitar Culture Industrial Park.

Zheng'an isn't making the best guitars in the world. For the most part, it's serving the lower half of the market. But its manufacturers are improving as local brands are getting hungry for global recognition. One of them is experimenting by adding bamboo into its guitars. Many of them are trying to become known for quality, not cheapness. I suspect many of them will get there. Chinese manufacturers are steadily gaining recognition for producing quality knives, sound systems, electric vehicles, consumer drones, and many other products. Why not guitars too?

After four days of cycling through Guizhou, we arrived in the municipality of Chongqing. The city's downtown core is built around two rivers—the Yangtze and the Jialing—with a skyline dominated by tall buildings that sprout from steep hillsides. They seem to stack upon each other: You can enter a building at ground level, go up by elevator for over ten floors, and exit once more at ground level. Chongqing is my favorite Chinese city to visit because it has the country's, and perhaps the world's, most dramatic urban setting. Highways and bridges weave through huge buildings that look as if they are carved into the hills, connected to each other by systems of stairs, escalators, and walkways. The city is filled with ludicrous designs, like a subway line that passes through the middle of an apartment building sitting on a hill.

Chongqing was China's capital during World War II, then known as Chungking. There, Chiang Kai-shek's Nationalist forces huddled with Communists and the US general Joseph Stilwell inside air-raid tunnels carved into hillsides to shelter from Japanese bombers. Chongqing is a municipality that matches the landmass of Austria—and is just as mountainous—as well as the population of Texas—and is just as boisterous. The bridges that were elegant in the Guizhou countryside swelled toward monstrosity as we approached the city. Everything is bigger in Chongqing. It is raucous, full of unexpected sites, a city that *teems*. With its *Blade Runner* aesthetic, Chongqing is the embodiment of cyberpunk—or more aptly given its rivers, hydropunk.

The mountains that protected the city from Japanese bombers also create a heat trap, making Chongqing one of China's "four furnaces." Perversely, the favorite food of locals is a cauldron of red chilies, beef oil, and Sichuan peppercorns—which generates a purring tingle on the tongue—into which one dips a swirl of thinly sliced meats and vegetables. Some of the air-raid tunnels have become hotpot restaurants, popular because the tunnels' cool air helps spicy food go down more easily. Chongqing is also making plans to turn some of these shelters into art exhibitions or wine cellars.

Christian, Teng, and I were in a celebratory mood when we reached the city. After four days of cycling through nature, it felt great to be thrown into Chongqing's dramatic urban scenery. At night, the city's skyscrapers come to life with bright lights dancing along their sides. As we watched the sun set, people gathered around low tables, the centers of which held steaming pots of crimson broth.

I almost never drink. If there was ever an occasion, I decided, it would be the end of this bike ride. The three of us toasted each other with cold beers and then ordered food so spicy that it altered my auditory capacity. Below us were lazy pleasure boats cruising on the Yangtze River, a few headed toward the Three Gorges Dam. The following day, I hopped back on the high-speed rail to return to work in Shanghai.

. . .

IT WAS ONLY AFTERWARD that I started to appreciate the strangeness of what I had cycled through. I had traversed a poor region to which the engineering state has devoted tremendous resources to modernize. Guizhou had compressed the century's worth of investments that the United States had made—between the Transcontinental Railroad and the Interstate Highway System—into two decades.

After cycling through Guizhou, I came to a different understanding of the term "socialism with Chinese characteristics."

China does little by way of redistribution from the wealthy to the poor; rather, it is enacting a Leninist agenda in which the state retains enormous discretion to command economic resources in order to maintain political control and to build toward a post-scarcity world. By examining Guizhou's development, as well as the developments of a few other places that I want to bring readers' attention to, we can grasp just what those "Chinese characteristics" actually entail.

Guizhou has built forty-five of the world's one hundred highest bridges. It has eleven airports, with three more under construction. It has five thousand miles of expressways, ranked fourth among provinces in China by length. It has around a thousand miles of high-speed train track. Guizhou's infrastructure isn't made only of the twentieth-century stuff of steel and concrete. Guiyang bills itself as a "big data valley," touting that its cool air can lower heating costs. Enormous facilities housing data servers make Guizhou emblematic of the modern infrastructure that powers AI too.

The Guizhou locals we chatted with were prouder of their bridges than anything else. My friends and I cycled across bridges that were set above plunging ravines. State media boasts that Guizhou has become a "museum of bridges," a few of which are trying to develop into tourism sites: The tenth-highest bridge in Guizhou (which is twenty-third globally) hosts the world's highest bungee jump. Each time the engineers build a bridge, they inevitably announce that travel times between two towns have been cut from many hours to perhaps a few minutes. That creates real convenience and connection for rural people. Some of these are bridges to nowhere, but after a few years, they become somewhere.

Still, beneath Guizhou's engineering marvels are counties mired in poverty. At $8,000 per capita, the province has the income of Botswana, 40 percent below China's national average and less than a third that of rich coastal cities like Beijing and Shanghai. One day, Christian remarked on how few working-age adults we saw in Guizhou: Those who don't have a job making guitars have mostly migrated to other provinces, leaving small children in the care of grandparents. In 2010, only half of Guizhou's children attended high school—the lowest rate in the country. News reports often featured stories of children having to rise at the crack of dawn and hike through harrowing mountain paths, some with rope ladders, to be able to attend school.

In spite of the challenges of deep rural isolation, China's fourth-poorest province—where household income is one-fifteenth that of New York State—has vastly superior infrastructure: three times the length of New York's highways, as well as a functional high-speed rail network. And Guizhou isn't exactly an exceptional Chinese province. Across the country, the engineering state has relentlessly built public works, making Guizhou an extreme case of China's growth strategy rather than a deviation from it.

Modern China has been on a building spree. It began in the 1990s, after economic reopening took hold, and then received another boost in 2008, when the central government approved vast public works to respond to the global financial crisis.

China's first interprovincial expressway opened in 1993, connecting Beijing with the nearby port city of Tianjin. Soon enough, highways reached everywhere. A Chinese citizen born when the country completed its first expressway would—by the time she reached the legal driving age of eighteen in 2011—be able to drive on a highway system that surpassed the length of the US interstate system. By 2020, China had built a second batch of expressways that again totaled the length of the US system. The first expanse of highways took eighteen years to build; the second took half that time.

Cars quickly filled these roads. In 1990, there were half a million automobiles in the country; in 2024, there were 435 million, many of them electric. China didn't just build cars and highways. It also built mass transit. From 2003 to 2013, Shanghai added as much subway track as in the entire system in New York City. In 2025, fifty-one Chinese cities have subway lines, eleven of which are longer than New York's. China now has a longer high-speed rail network than the rest of the world put together, ten times the length of Spain's and Japan's (second and third in the world, respectively). Sleek railcars in silver zooming on elevated bridges are telegenic things, their pictures adorning billboards and book covers. This system completes around two billion passenger trips each year.

The state loves showing footage of big container ships that berth under enormous cranes, plucking from a mosaic of containers. As exports soared, China's ports became the world's busiest. Shanghai alone moved more containers in 2022 than all of the US ports combined. China's export engine sputtered in the early 2000s, not for a lack of ports but for a lack of power in Guangdong. So the state invested in a network of new power plants mostly burning coal. In addition to using fossil fuels, China builds a third to a half of the world's new wind and solar capacity each year. It is sending renewable energy from its sparse western provinces into its industrialized eastern provinces.

In 1957, the world's first commercial nuclear plant started producing electricity in Pennsylvania. In 1991, China's first commercial nuclear power plant started producing electricity. By 2025, China caught up to the United States in the number of nuclear plants: fifty-five and fifty-four, respectively. Though the United States might restart a few decommissioned reactors, it has just one under construction. Meanwhile, thirty-one are under construction in China. The only US nuclear plant built in the twenty-first century took fifteen years and $30 billion. In August 2024, China's nuclear authority approved construction of eleven new reactors, which are collectively expected to cost the same amount.

Above all, China built housing. Its urban population has grown by an average of sixteen million people each year since 1978, which means, in effect, that the state built a new city the size of greater New York City and greater Boston combined every year for thirty-five years. Though Beijing, Shanghai, and Shenzhen have soaring housing prices, high rates of construction plus rising wages have broadly improved affordability. From 2007 to 2018, the average price of an urban apartment fell from nine times the average household income to seven times. This building spree consumes colossal amounts of steel, aluminum, copper, cement, and glass. According to Vaclav Smil, the 4.4 billion tons of cement that China produced from 2018 to 2019 nearly equals the amount of cement the United States produced over the entire twentieth century.

This building boom was both a cause and an effect of China's growing wealth. It stimulated economic activity directly: The construction of homes, highways, subways, and power plants spurred demand for materials and jobs that rippled beyond the immediate construction site. It also facilitated China's urbanization, pulling people from farms into cities, where their productivity was much greater. In a crucial period while China's labor force was expanding, this infrastructure laid the foundation for the country's export-based manufacturing strategy.

Most of China's enrichment has been driven by the people themselves, finally freed from Maoist restraints to pursue a better life. Meanwhile, the state's mania for building public works has helped the country grow faster. You can see how China differs from India, Indonesia, and other developing countries, where growth is lower in part because the state hasn't built enough housing and infrastructure for their citizens.

While China compressed more than a century's worth of American construction into a few decades, it folded in many of its problems too.

Highways have ripped apart too many cities in China, just as they have in the United States. Chinese have mustered tremendous enthusiasm for destroying the nation's physical heritage in the recent past. It was prominent during the Cultural Revolution, when Mao ordered Red Guards to loot Buddhist temples, smash Confucian statues, and desecrate ancestral tombs. Over more recent decades, destruction was more systematic than the ruin of particular cultural treasures, as whole neighborhoods fell to the bulldozer. In their place are wide avenues and concrete superblocks. Unfortunately, not much new construction in China is optimized for charm and beauty.

The engineering state is built for a bird's-eye view. The geometry of highway interchanges, rows upon rows of solar photovoltaic panels, and, under the right lighting, even a belching chemical plant can produce a pleasing thrill when viewed up high and at a distance. Down below, the urban environment is not always pleasantly livable. Big cities like Beijing and Shenzhen are poorly laid out, with no extensive walkable zones. It takes forever to get across town.

I was much happier to live in Shanghai, where many streets have remained human-scaled rather than being built for cars. The French Concession, where I lived, remains leafy and full of cafés. Shanghai is highly walkable, and one is rarely more than a fifteen-minute walk from one of the city's many subway stations. Shanghai has vowed to open 120 new parks every year until 2025, when the city will reach 1,000 green spaces. The city of twenty-five million people works remarkably well. Like Tokyo, it has flourishing spaces for commerce, where little dumpling shops are tucked away even in subway stations. And Shanghai is superbly connected by high-speed rail to nearby cities—for example, Hangzhou, home to tech companies like Alibaba, and Suzhou, where many multinationals have manufacturing operations—which are themselves some of China's most successful cities.

Though China has embraced American car culture, it's still easy to get around by bike in Shanghai. The city has in recent years refashioned a stretch of its riverside into a series of wetland parks along a fifteen-mile bike path, where one can cycle past the brick warehouses and glass skyscrapers that make Shanghai feel quite like New York City. I loved taking my Giant along the river, zooming past the World Expo development, the Mercedes-Benz Arena, bridges tall enough to allow barges to sail underneath, and all sorts of beautifully preserved industrial buildings.

Compulsive construction has benefits. Though people in Guizhou remain poor, the villagers we encountered on our bike trip told us they are thrilled to have new bridges and trains. For Chinese who have experienced economic growth rates of 10 percent a year, it would feel like their country was reborn roughly every seven years. (That's how much time an economy takes to double with that growth rate.) It means better cars, more subway lines, cleaner streets, more parks, and a hundred other improvements.

• • •

THE UNITED STATES USED to make enormous investments to modernize its poorer regions. Today, Americans rarely feel so excited for major construction projects, in part because they're associated with environmental damage, in part because they take so long to complete, and in part because they're so rare that people have forgotten how much they can improve lives.

Americans are no longer able to appreciate that a physically dynamic landscape creates a sense of progress. People living in Texas,

Arizona, and the southern states that have built new skylines and masses of new homes might know how that feels. But in the largest cities in the Northeast and California, the default is toward rigidity. A new building here and there, perhaps a cute new shop or café, a toilet that cost over a million dollars—these overall inspire little eagerness for physical change.

That feeling contributes to a blind spot that Americans have for China. People unable to appreciate the benefits of material improvements also don't understand how it produces pride and satisfaction. China's transformation has given people running water and toilets, mass transit and highways, beautiful parks and modern malls. Most people can remember a time in only the recent past when they didn't have these things. This growth trendline matters. The glittering skyscrapers and rail lines form a core plank of the Communist Party's legitimacy. Though China's growth has slowed substantially under Xi's rule, people have a hope for improvement. The better infrastructure that has been built helps people to feel that progress still courses throughout the country.

When Beijing began construction of its high-speed rail program in 2008, critics charged that it was foolish for a then-poor country to acquire the sorts of luxury infrastructure out of reach even for many rich countries. "Infrastructure investment can be too good for a country's development level," concluded a line from economist Michael Pettis, which was not an atypical sentiment. But China's railways had been hugely crowded, with passenger trains sharing the same tracks as freight trains, which caused endless delays. The creation of this fast and dedicated passenger network relieved congestion for all.

A study undertaken by the World Bank in 2019 found that China's high-speed rail system is economically viable, with ticket revenues able to recoup costs. China has been able to build high-speed rail cheaply because it has standardized designs and excellent project management. The average cost to construct a high-speed line in China is about $33 million per mile, which is 40 percent cheaper than in Europe and 80 percent cheaper than California's effort, which has seen costs balloon to $192 million per mile. By taking a broader view, the World Bank suggests that China's high-speed rail has delivered substantial benefits beyond ticket revenues, which includes saving time for users, increasing intellectual and business exchanges, reducing road accidents and traffic congestion, and lowering carbon emissions.

Rather than redistribute resources from the rich to the poor, the state builds infrastructure in Guizhou. Lenin used the term "commanding heights of the economy" to refer to strategic sectors like power generation and transportation. In Guizhou, the commanding heights may well be seen from its tall bridges.

The engineering state, citing socialism with Chinese characteristics, is set up to give people one main thing: material improvements, mostly through public works. The engineering state builds big in part because it's made up of self-professed communists who grew up admiring the Soviet Union. Communist Party leaders like Xi Jinping studied in an educational system steeped in Marxism. For them, production was a noble deed to advance communism, while consumption was a despicable act of capitalism. This party believes that only the state has the wisdom to invest in strategic megaprojects, whereas consumers will waste money on themselves. It is hostile to ordinary people having much command of resources, which empowers an individual's agency rather than the state's.

The Communist Party celebrates the birthday of Karl Marx; to close out its party congress, held twice a decade, the military band plays the "Internationale" socialist anthem inside the Great Hall of the People. But as I said in my introduction, China is also a country governed by conservatives who masquerade as leftists. Perhaps no other self-proclaimed socialist country is as lightly taxed as China. Nearly three-quarters of China's population are spared from paying income tax. China has also failed to levy a broad property tax, leaving the bulk of rich city dwellers' wealth untouched. It relies more heavily on consumption taxes, which are regressive because they burden the poor more than the rich.

Beijing has announced several times that it would impose a property tax. Each time it faltered. One of the political reasons is that China's leaders are familiar with the American slogan "No taxation without representation." Since the state levies relatively light taxes, which it takes unobtrusively from citizens, it reduces the risk that people start asking questions about what the state is doing with their hard-earned funds and whether their taxes should entitle them to greater political participation.

Low taxes make China stingy on welfare. Around 10 percent of its GDP goes toward social spending, compared to 20 percent in the United States and 30 percent among the more generous European states. China's pension and health care spending are much lower than that of other rich countries. It is especially miserly with unemployment insurance: Only about a tenth of China's unemployed are eligible for modest benefits. Occasionally, Chinese leftists have protested the state of affairs. Rather than provide better welfare in response, the state has detained students trying to organize Marxist reading groups.

Xi has forcefully pushed back on the idea that China needs more generous welfare. In a major speech in 2021, he said, "Even when we have reached a higher level of development . . . we should not go overboard with social transfers. For we must avoid letting people get lazy from their sense of entitlement to welfare." Worrying that welfare could make the people lazy is one of those instances when a Communist Party leader sounds like Ronald Reagan.

China's economic model isn't a simple-minded application of Marxism. The Communist Party would say that its system is modulated by certain Chinese characteristics. The sort of central planning that is part of Marxist-Leninist states have certain resonances with the centuries-old predispositions of China's engineering state, especially construction and control. But China has some element of capitalism as well, which explains why the country has created a far more durable economic model than the failed Soviet-style states.

Construction, capitalism, and control. These elements are sometimes in tension. After China's digital platforms grew powerful and profitable, the Communist Party reined them in (the focus of my sixth chapter). It found a lot to dislike among tech tycoons and their business models. Companies and people were engaging in transactions—buying goods, borrowing money, contracting for services—without mediation by the state. And digital platforms created billionaires who could not resist flaunting their wealth or wisdom, much as their Silicon Valley counterparts do. Subsequently, the Communist Party smashed many of their businesses before they had begun to wield real power. The state wants to have the ultimate say in controlling economic relations throughout society.

The lack of a safety net is one of the reasons that Chinese households save a great deal of their income for contingencies. The engineering state likes that just fine. Xi's generation came of age in the 1950s, when China imitated a Stalinist program for intensive control over enterprises as well as a focus on industrialization and heavy industry. If people can bear some pain now and save, the Communist Party has long been saying, then life in the future will be better.

Xi was born in 1953, the year that Beijing unveiled its First Five-Year Plan, which concentrated the state's resources to build seven hundred industrial projects. Drawing directly from Soviet practices, it also had substantial aid from the Soviet Union, which provided technical guidance to projects that included metallurgy plants, chemical facilities, and defense projects. In 2020, Xi announced the Fourteenth Five-Year Plan, the ambitions of which are more breathtaking than anything the Soviets attempted.

The engineering state isn't finished with building big. "We will perform basic scientific research on the origin and evolution of the universe, carry out interstellar exploration such as Mars orbiting and asteroid inspection," goes the opening section of the Fourteenth Five-Year Plan on science and technology. It gets better from there. "We will construct hard X-ray free-electron laser devices, high-altitude cosmic ray observation stations, comprehensive extreme condition experimental devices, deep underground cutting-edge physical experimental facilities with very low background radiation." China wants not only to explore deep space but also to use "heavy icebreakers" for polar exploration in the deep sea.

"We will add 3,000 kilometers of urban rail transit" states the section on mass transit. The plan specifies the sections of highways and high-speed rail that it will build. It has major targets for energy: "We will build hydropower bases on the lower reaches of the Yarlung Tsangpo River," which will have triple the power-generating capacity of the Three Gorges Dam, and the construction of ultra-high voltage transmission lines to connect power from the country's west to east. It has a plan for climate change, especially water management. Beijing will work on the South-to-North Water Diversion Project, which feels like a throwback to the Grand Canal of the seventh century AD. It involves a gigantic effort to draw water from China's southern rivers toward its parched northern cities, along three canal systems, targeting completion in 2050. The plan envisions the creation of large water reservoirs across the country and the construction of major flood-control projects.

The Fourteenth Five-Year Plan outlines interstellar research and other state-directed megaprojects. There's something for the ordinary consumer too, but it's nowhere near as exciting. To promote consumption, the plan suggests measures like "expanding the coverage of e-commerce in rural areas," "improving product recalls," and "improving in-city duty-free shops." Fine measures, but puny relative to orbiting Mars. The economic planners have obviously poured their hearts into the scientific projects, whereas the consumption measures look like a hasty afterthought. When Chinese officials talk about promoting consumption, it often involves building new malls or replacing old industrial equipment. In other words, it's still more about investing to build stuff rather than shifting the propensity of households to spend a greater share of their income.

Under Mao, China practiced a more literal form of Marxism, with full state control of the means of production. Deng Xiaoping pivoted the country away from that failed experiment. As Deng was fond of remarking, the defining feature of socialism was not economic redistribution but rather "concentrating resources to accomplish great tasks." That flexible definition allowed for greater adaptability, generated higher growth, and sustained the regime into the twenty-first century. Under Deng's definition, the United States has also achieved plenty of socialism. The Manhattan Project, the Interstate Highway System, and the Apollo Program all concentrated resources to accomplish great tasks. Maybe even Reagan's Strategic Defense Initiative could have been understood as socialism.

. . .

WHEN THE ENGINEERING STATE works, it can produce beautiful cities like Shanghai. But Shanghai is exceptional: It has been China's richest and most westernized city for the better part of a century. The engineering state also produces a lot of problems. To see them, we should return one more time to Guizhou.

Under the gleaming new bridges lurk not only poverty but also a massive debt burden. The underlying hope of Guizhou's construction is that infrastructure will invite lasting economic activity. Part of that has worked out: Guizhou incomes have risen by nearly 10 percent annually from 2011 to 2022, driven partially by urbanization and by the tourism facilitated by new infrastructure.

But most of Guizhou's infrastructure spending looks dubious. Its super-high bridges aren't producing the revenue to recoup anywhere near their super-high costs. Of Guizhou's eleven airports, five have less than a dozen flights each week—and there are three more airports still under construction. Guizhou has become one of China's most indebted provinces, and it's starting to feel real fiscal distress. In an unusual move, Guiyang's finance bureau issued a public outcry in 2022 that it was at the end of its ability to deal with the debt. Quickly afterward, the government deleted its own admission.

Guizhou's debt has kindled Beijing's wrath. In China, the only people scarier than debt collectors are political inspectors from the central government. The Communist Party has unleashed teams of officers from the Central Commission for Discipline Inspection to descend on Guizhou. They are unbound by even the modest levels of legal niceties afforded in China. Rather than investigating legal crimes, their remit is to find "violations of party discipline," a nebulous charge that includes not only corruption but also misuse of public funds and political disloyalty to the Communist Party. That makes the commission akin to the Inquisition, enforcing doctrine and discipline on its members.

Financial inquisitors found something in Liupanshui, the westernmost city in Guizhou, home to the world's highest bridge. Li Zaiyong, sixty-two, was a handsome man who had big plans for his city. In the three years that Li was party secretary of Liupanshui, he authorized twenty-three tourism projects, including elegant Chinese temples and replicas of European town squares, which looked pretty from a distance but poorly painted up close. Li aimed to transform his city into a ski town, though Liupanshui would be lucky to get a few inches of snow each year. To attract skiers, he built a cable lift that he declared to be Asia's longest, as well as dozens of artificial snowmakers to spray powder on bunny slopes. And Li blanketed the hillsides with orchards of chestnut rose hips, a bulbous yellow fruit bristling with spiny protuberances, which looks slightly nightmarish but is much loved by locals for its sweet flesh.

With enough will and snowmakers, Li believed he could create a tourism hub from scratch.

Liupanshui locals greeted Li's plans mostly with skepticism. Though their region is home to waterfalls, karst caves, and stunning green mountains, the city itself has little of beauty. Local industry consisted of coal and iron mining. One man wondered to a TV crew, "We don't have much to see here. How much money would we have to spend to create something worthwhile?" Li Zaiyong believed a great deal, and he mobilized a great deal of funds to make things happen. Since he was the top official in the city, local banks had a hard time saying no to him.

But none of Li's efforts bore fruit.

Liupanshui never developed into an appealing ski destination: China's skiers went to the northeast in the winter, which has real slopes and real snow. Richer tourists from Beijing and Shanghai skipped Li's gaudy European facsimiles for the real deal in Venice and Vienna. The faux European town squares have been taken over by local black goats, which treat the lawns as grazing grounds. Even the chestnut rose bushes died.

All that the city got for its troubles was $21 billion of new debt, an enormous amount for a poor city in a poor province. The Central Commission for Discipline Inspection denounced Li's investments as "vanity projects" and hustled him into its secretive judicial proceedings. In 2024, state media made an example of Li in a primetime documentary. He still had good looks in prison. But detention had turned his hair gray, whether from stress or lack of access to hair dye. As Li spoke inside a dimly lit room, he explained his reckless spending: "It was the nation's money, not mine."

Li misused public funds. But he was also playing a political game recognizable to any other party secretary. One of the Communist Party's personnel practices (inherited from imperial times) is to rotate officials between various jurisdictions, forcing them to gain broad experience and preventing them from drawing their power base from their home province. China has few officials with careers like Joe Biden's, who, before becoming vice president and then president, spent his entire political life representing Delaware. Li Zaiyong had been an official all over Guizhou before landing in Liupanshui. The way for him to reach even higher office was to demonstrate a track record for growth.

The political system of the engineering state rewards construction. China's political leaders, after all, are selected, not elected. To reach higher office, they are given probing assessments by the Organization Department, which, along with the Propaganda Department and the Central Commission for Discipline Inspection, are the Communist Party's most important governing instruments. The Organization Department evaluates local leaders on a few soft metrics, like leadership, loyalty, and resistance to graft. The department also assessed whether an official was capable of stimulating economic growth while suppressing political dissent.

But few party secretaries have great ideas for growing the township or province they've landed in. Furthermore, since local governments don't have property taxes, they primarily fund themselves through land sales to real estate developers. This combination of personnel policy and fiscal quirks produces officials like Li Zaiyong who invest in glamorous projects and whose failures are apparent only after they've left office.

Li's larger goal was to impress his superiors, and on that he succeeded for a while. The party promoted him to be a vice governor of Guizhou, a position he held for five years before his fall. When the debt came due, Li's career was over. A few months after his televised confession, a provincial court sentenced Li to death with a two-year reprieve.

Li wasn't the only Guizhou official to be detained. Beijing investigated a huge number of mid- to high-level provincial officials in 2023, ensnaring even a former party secretary—Guizhou's highest official. Again, the situation in Guizhou was not an aberration from China's growth strategy. Dubious projects can be found all over the country, since nearly every province has laughable replicas of European town squares that fail to attract tourists, underused infrastructure that can't repay bondholders, and overbuilt cities that struggle to chart a course away from resource extraction.

Li dreamed up crackpot schemes for development. It would have been what any party official would do if they were dropped into a land as unpromising as western Guizhou. But he had grown too brazen in a political game very much designed by the central government in Beijing.

◆ ◆ ◆

THE ENGINEERING STATE IS not all glittering Shanghai. It is sometimes Liupanshui, a city that made all the wrong investments. Sometimes it is Tianjin, which was a success until it overbuilt. When I lived in Beijing, I heard the name of nearby Tianjin used as a byword for excess. One day, I took a thirty-minute train ride from Beijing to see it for myself.

Tianjin is relatively rich. In the 2000s, it spent heavily to build a financial district, styling itself "China's Manhattan." It was a fanciful branding exercise, though the city did manage to import a real institution from Manhattan: The Juilliard School opened its first international campus there in 2020. Tianjin used to be among China's most industrialized cities—when the country looked to the Soviet Union for all its economic ideas. Now it is emblematic of China's rust belt and nothing like a real financial zone. The skyscrapers in Binhai, the zone that Tianjin has designated a financial district, are mostly empty. On the weekday that I visited in 2020, the central mall in Binhai had a few people walking around, but almost none of the commercial buildings looked occupied. China's Manhattan was hollow.

Tianjin has built not only China's third-tallest skyscraper (ninety-seven floors, little occupied) but also one of its most photogenic libraries. The Dutch architects behind Tianjin's Binhai Library put a bright white sphere in its center, around which undulating curves make up shelving space. Except few of these shelves held any books. Once I got up close, I could see that the beautiful shelves had only digital prints of book spines. All around me were people taking selfies rather than browsing or reading.

I sometimes think of Tianjin's library as a metaphor for China's economy: great hardware that looks impressive from a distance, not filled with the softer stuff that actually matters. Tianjin could have focused on filling its amazing skyscrapers with better businesses. Instead, it could only build more hollow shells, while it gained considerable debt.

Moody's, the American credit rating firm, listed Tianjin and Guizhou as China's two most heavily indebted regions. Each has a debt-to-GDP ratio approaching that of Italy's. In 2018, Tianjin acknowledged that Binhai's growth was far overstated, forcing it to revise down its GDP by nearly 20 percent. It is rare for the Chinese government to acknowledge data fraud. So it's all the more curious that the central government is trying to pull off another major development scheme nearby. Not far from Tianjin is one of Xi Jinping's signature new initiatives: the Xiong'an New Area. Xi has declared that Xiong'an will be the country's most modern city and central to China's "strategy for the next millennium." Xiong'an is likely to receive extensive new investments because Xi has given it so much personal attention—unless that attention wanders, or if Xi is no longer around, in which case Xiong'an could turn into another Binhai.

The creation of urban megahubs, like the zone connecting Beijing to Tianjin and Xiong'an, is a part of a bet. The central government has designated a dozen urban regions for concentrated investments. The five largest—Beijing in the north, Shanghai-Hangzhou-Suzhou in the east,

Shenzhen-Hong Kong-Guangzhou in the south, Wuhan-Changsha in central China, and Chongqing-Chengdu in the west—average 110 million people, each nearly the population of Japan. The state is investing in rail, subways, buses, and highways to knit them into regional hubs. Alain Bertaud, a former principal urban planner at the World Bank, has told the *Economist* how these agglomerations can achieve previously unseen levels of productivity, representing the difference between England and the rest of the world during the Industrial Revolution.

It's part of a growth strategy that involves building a lot of stuff. That's not something that only China's government does. Its companies do it too. The United States and Europe have launched trade wars on the basis of the overproduction by Chinese firms, lodging diplomatic pro-tests and countervailing tariffs against steel, aluminum, solar photovol-taic panels, and electric vehicles. The engineering state is much more interested in promoting building and manufacturing than services.

China now has the capacity to produce around sixty million cars a year (one-third electric, two-thirds combustion), out of an annual global market of around ninety million cars sold. China's domestic market absorbs less than half of its production. China produces so much in part because every province wants to be an automotive manu-facturing hub. The country has over a hundred automotive brands, most of them small, all of them fighting for sales. The competition is so fierce in part because auto companies receive extensive support from local governments, who all try to promote *their* champion through cheap credit and consumer rebates to local companies. Shanghai, for example, is full of the locally produced SAIC-Volkswagen cars, while Shenzhen is dominated by its own champion, BYD.

Sometimes not even bankruptcy can stop an automaker from production. Zhido, a producer of small EVs, went bust in 2019; five years later, it had restructured, with government help, and restarted its production lines. NIO Inc. was right on the brink of bankruptcy in 2020 until its home government of Hefei rescued it; the company has since turned around its fortunes and is once more shipping its electric vehicles. The United States offered extraordinary bailouts to Detroit automakers in the aftermath of the financial crisis. In China, local governments help companies out every day. As a result, few of the brands can achieve really big scale, and China has to depend on exports to absorb all the vehicles domestic consumers aren't buying.

China's government is much more focused on the smooth functioning of the supply side of the economy rather than on helping consumers. This principle became really apparent in 2020. While Western governments reacted to the global pandemic by sending cash payments to households—in the United States, three rounds totaling $3,200—Beijing gave little financial support to people. It offered only meager increases to unemployment insurance, which only a fraction of the millions of unemployed workers were able to claim.

Beijing decided instead that the best way it could help workers was to get them back to work. In practice, it meant assisting companies to restart production rather than sending cash to households. It did that with its pursuit of the zero-Covid strategy, which involved taking drastic action to seal the country off from foreign travelers and imposing protracted lockdowns wherever it detected the SARS-CoV-2 virus. Beijing wanted especially for manufacturers to maintain operation. While the rest of the world was having trouble producing stuff—masks and cotton swabs, electronics for remote work—and while foreign consumers had stimulus money to spend, Chinese factories were positioned to meet the world's demand.

For a while, that scheme worked. China's trade surplus hit a record high in 2021, and then again in 2022, approaching almost a trillion dollars. In spite of the tariffs that Donald Trump imposed on gigantic categories of Chinese-made goods, the United States and China experienced record trade in 2022. China's manufacturing surge, however, could not make up for the losses created by lockdowns elsewhere in the economy. But Beijing's attitude was that so long as manufacturers could crank out more goods, then the economy was robust and there would be little need to hand out checks or any other sort of welfare.

And the manufacturers were cranking out a lot of goods. In the fall of 2020, I remember visiting the factory of a technology manufacturer outside of Shanghai. An executive had invited me in to tour his new production line. He was a Chinese national who had mostly worked for American companies and still traveled between the two countries. As we sipped tea in his office after the tour, we chatted about why the United States was then mired in production difficulties, unable to make much of the personal protective equipment that people wanted. "American manufacturers constantly asked themselves whether making masks and cotton swabs was part of their 'core competence.' Most of them decided not." He put down his teacup and looked at me. "Chinese companies decided that *making money* is their core competence, therefore they go and make masks, or whatever else the market needs."

In 2020, I could have picked up face masks that were branded Foxconn (the world's largest electronics contract manufacturer), BYD (the world's largest electric vehicle manufacturer), or JD.com (China's second-largest e-commerce platform). Companies retooled some of their production lines to get into the masks and money business. Chinese conglomerates rarely hesitate to go after the core business lines of others. Huawei, for example, expanded from making telecommunications infrastructure equipment to tread on companies like Xiaomi, which makes smartphones. And both have now leapt into the automotive business. This sort of expansion is driven both by the fiercely competitive market environments and by government subsidies that make it easier for companies to try their hand at making new products.

They allow companies to unleash a flood of undifferentiated products, ruthlessly underbid each other, and pray their competitors run out of money before they do. China now dominates the solar industry, but almost no firms are happy because of the overcapacity. Many of these Chinese companies will inevitably go out of business, after they've dragged down their competitors all over the world in brutal price wars. This trend has produced a frustrating quirk in China's equity markets. Financial investors have seen that there is no relationship between Chinese stock market performance and GDP growth. Although the economy has grown by a factor of eight in real terms between 1992 and 2018, the Shanghai Composite Index has been one of the worst-performing major indices. In China, for a variety of reasons that includes weak corporate governance, onshore stocks dance to their own tune. Part of the reason is that even for technologies that Chinese firms dominate—like solar photovoltaic panels—few firms are able to make much profit.

·   ·   ·

FINANCIAL INVESTORS HAVE NO need for our sympathy. There are far bigger victims of socialism with Chinese characteristics.

The environment is a prominent victim of all this construction. China's environmental reviews are not unserious, but they are almost always subordinate to economic development. All those highways are made from hulking amounts of steel and concrete, which have pulverized many habitats. Their construction also requires enormous amounts of energy: China now burns more coal than the rest of the world combined. Though the country's air quality has improved over the past decade, this obsession with heavy industry is why a gray and dreadful smog still descends on many of its cities.

China doesn't seek to protect the environment. It tries to engineer away the problems. Over the past five years, the country has been repeatedly struck by climate-related tragedies, and it is hard to prove that the investments China has made in flood control and water diversion megaprojects have improved matters. In the summer of 2022, a year after my bike trip to Chongqing, I returned to find the city in a historic drought. I was stunned to see that one of its two rivers, the

Jialing, had nearly run dry; even the mighty Yangtze had prominent dry patches. People tried to avoid heat by staying indoors. But many of them couldn't turn on air conditioning because the rivers had lost so much flow that even the hydropower failed.

China's other climate disasters have included major floods in Henan province (in which fourteen people drowned in a subway train, according to official numbers), power outages in central China over the winter of 2022, and floods that displaced more than a hundred thousand people in Guangdong in 2024. Perhaps megaprojects have ameliorated what were already disastrous conditions; perhaps they had no impact at all. But environmental scientists do often question whether this sort of engineering has made things worse. The construction of a dam might well provide flood relief, or it could worsen a drought by reducing downstream flow and increasing evaporation losses.

When the state builds big dams, it floods ecosystems and dispossesses residents. The world's biggest dam is the Three Gorges Dam, not far from Chongqing. Building it has demanded the resettlement of up to 1.5 million people. The government's compensation for resettlement is often generous. But it has limited patience for suffering residents who block development, whether that's a new highway or a new mall, eventually moving holdouts by hook or by crook.

The worst-affected people are targeted minority groups, who have to bear Beijing's social engineering. The state has singled out, for example, Tibetans, who are forced to relocate from high-altitude mountains, where they are able to graze their yaks and horses, to lower-altitude farms in part to monitor them more easily. What are yak herders supposed to do when they move down to apartment blocks? Rural people who know only their farming or pastoralist lives are often at loose ends when the government resettles them into rows upon rows of high-rises. Two researchers at the University of Colorado have documented China's coercive tactics to compel locals to leave their homes. It is a process it calls "thought work," ranging from presenting resettlement as a voluntary and happy choice to holding intensive one-on-one meetings with recalcitrant folks who do not want to leave. Officials mix inducements with threats until they wear down the farmers. Thus, the state has been able to achieve "voluntary" resettlement rates of 100 percent.

Reckless construction has often produced rubbish quality. Builders employed cheap materials to construct even schoolhouses. The 2008 earthquake that tore through Sichuan also shattered thousands of schoolrooms, killing five thousand children (according to official figures). Grieving parents who tried to take investigations of official corruption into their own hands have faced detention. Public works give government officials plenty of discretion about how to build a project, giving them a lot of opportunities to accept kickbacks. Even if officials are upstanding, the developer might contract out the construction to a lower-cost builder, who takes a margin and subcontracts out again, and on and on until it reaches someone willing to cut costs to the bone. Parents called the collapsed schools in Sichuan "tofu houses" for their fragility. Building big, in other words, does not always mean building well.

Many construction projects represent a tremendous waste of the steel and cement that was produced by burning so much coal. There are better uses for these resources: softer concerns around health and education, not the gigantic hardware of more highways.

Though rich students in Shanghai score splendidly on international exams, education in China's rural areas is still often abysmal. The Covid pandemic revealed that the country's health care system is weak, with shortages of doctors and nurses and six times fewer intensive care unit beds per capita than in the United States. An official like Li Zaiyong might be more interested in building a gleaming hospital filled with sophisticated equipment. Their attention drifts, however, when it comes to installing the trained technicians capable of operating the facility, since the Communist Party is better at rewarding new construction than health outcomes. The engineering state is focused mostly on monumentalism. Though there are many public toilets, provision of toilet paper is only a sometimes thing. Nowhere in China is it advisable to drink tap water. Not even Shanghai.

The engineering state has engaged in wild spasms of building over the past four decades. That has achieved considerable wonders and a fair degree of harm. The future would be better if China could learn to build less, while the United States learns to build more.

I've come to realize that there are many ways that China and the United States are inversions of each other. Households save a great deal of their earnings in China, while it is really easy to borrow money or spend on credit in America. In terms of national policy, China is much more focused on the supply side of the economy: It suppresses consumption as it favors manufacturers with preferential financing and all manner of policy support. The United States, meanwhile, is focused on regulating demand, for example, by imposing rent control in expensive cities or mailing out checks to consumers during the pandemic.

Both approaches are running into problems. China won't become the world's biggest economy by building more tall bridges. It also can't continue manufacturing more than twice the number of cars it sells at home. And the United States is starting to realize the problems of being too focused on the demand side of the economy. When the federal government offers, for example, rental support in housing-scarce cities, landlords can raise their prices, leaving renters no better off. When it increases financial aid for spiraling college tuition costs, universities are able to eat part of that by raising their tuition. Under banners like "abundance agenda," "supply side progressivism," and "progress studies," various movements are trying to loosen American supply constraints. These are excellent ideas that I hope will be broadly adopted.

The economic partnership between the United States and China made many groups better off. But it also exacerbated the problems inherent in the economic systems in both countries. Overreliance on Chinese manufacturing accelerated US neglect of its supply side. Meanwhile, China hasn't felt the need to wean off its dependence on exports because American consumers are always there to buy its goods. As these countries grow apart, they are going to have to do something difficult: The United States will have to regain all the muscle it has lost for building public works as well as manufacturing capacity, and China will have to empower consumers by getting over its fear of making people lazy.

Doing these things won't be easy for either country. Any time the Chinese economy wobbles, Beijing's knee-jerk response is to announce another gigantic public works package. After a year of sluggish growth at the end of 2023, Beijing announced it would spend a cool one trillion renminbi (or $140 billion) on flood prevention and natural disaster resilience. Its instinct is still to keep building, as each of its Five-Year Plans reveal. Government ministries and state-owned enterprises are always formulating plans for the next rail extension, the next bridge, the next set of subway lines. Since the planning is already completed, a fresh infusion of funds can have a quick impact on growth, with spending on a new bridge making an impression on economic statistics immediately. Never mind that China has gotten less growth from each unit of new investment since its big infrastructure binge of 2008. The Communist Party continues to build because it's full of engineers and also because Marxist-Leninists don't want to cede economic agency to the people.

China would be better off if it built less and built better. But we should also resist judging it by the standards of the United States, which is frankly underprovisioned in public infrastructure. Because there is perhaps one thing worse than an overactive state that can't stop moving—and that is a state that can't move at all.

When I look at the United States, I marvel both at how much it did build before 1970 as well as how little it constructed afterward. China spent 13.5 percent of its GDP on infrastructure investment in 2016, whereas the US average over the past three decades is closer to 3 percent each year. Could not the two countries just move a few percentage points closer to each other?

I wrote this book mostly out of my office at the Yale Law School. New Haven is well connected to New York City on the Metro North trains, which are comfortable and reliable but a bit slow. One day, I came across a Metro North timetable from 1915. It revealed that the express train from New York's Grand Central Terminal to New Haven took the same amount of time then as in 2025: around two hours. The comparison isn't totally fair because trains today make more stops than before. But I think it is reasonable for Connecticut residents to demand faster service than what was available a century ago. The entire American Northeast badly needs better train service. At present, its only high-speed train (the Acela) would be stripped of that label if it operated anywhere in Europe or Asia.

One might think that it's not the end of the world for the United States to build gingerly and at extravagant cost; it is a rich country, after all. But slowness today risks global disaster. There is no way to achieve large-scale decarbonization without large-scale construction, of the sorts of solar, wind, and electrical transmission projects that China has been so good at.

Though the Biden administration committed enormous funds to address climate change, the country moves far too slowly on building things. One cautionary tale: the story of Cape Wind, the United States' first effort to develop offshore wind turbines. A developer tried to build turbines off the coast of Massachusetts, harnessing sea winds that are smoother and faster than those on land. Unfortunately, Cape Wind was in Nantucket Sound, home to some of the wealthiest, and mostly liberal, US citizens, like the Kennedy family, whose compound is in Hyannisport. These residents banded together, formed a nonprofit, and enlisted lawyers that included one of Harvard's best-known constitutional law professors to challenge the development. After sixteen years of lawsuits, the developer abandoned the project.

Environmental reviews continue to delay renewable projects. In 2024, the United States had 42 megawatts of operational offshore wind production, 932 megawatts under construction, and an astounding 20,978 megawatts undergoing permitting review, most of which are waiting on environmental analyses to be completed. Meanwhile, China is building most of the world's renewable energy. In 2023, while the United States added 6 gigawatts of new wind installations, China added 76. That year, China built two-thirds of the world's wind and solar plants, as well as four times more than the rest of the G-7 group of rich countries put together.

The lawyerly society is great at protecting the wealthy. The engineering state has a limited tolerance for how long infrastructure can be held up. It's barely imaginable that a group of wealthy people would be able to use legal means to force the cancellation of clean energy projects in China. If it is really going to be a climate emergency, then the rest of the world will need to move as fast as the engineering state.

Americans are starting to regain an awareness of the virtues of building. This political consciousness has budded in the political left, which has tended to favor physical stasis in the name of environmental protection or neighborhood preservation. Ezra Klein of the *New York Times* has pointed out that it's hardest to build in the most Democratic-leaning locales: high-speed rail in California, the Second Avenue subway in New York, and housing in practically all big cities. In *Abundance*, Klein and Derek Thompson advocate to unblock constraints and achieve supply-side progressivism.

Here is where socialism with Chinese characteristics can shine. Building big can sometimes unblock market power. People in Guizhou may not have much. But they do point to new bridges with pride, while using new roads and high-speed rail to get to markets and cities. Infrastructure that cannot recoup its revenue might upset bondholders and banks. But they represent subsidies to social benefits enjoyed by regular people.

Has China already been practicing supply-side progressivism? No, because nothing about it is "progressive" in a way that someone on the American left would understand. China's means of construction entail evicting people from their land, adopting a relatively lax approach to environmental protection and worker safety, and interpreting the public interest without substantial engagement with the actual people.

China's overbuilding has produced deep social, financial, and environmental costs. The United States has no need to emulate it uncritically. But the Chinese experience does offer political lessons for America. China has shown that financial constraints are less binding than they are cracked up to be. As John Maynard Keynes said, "Anything we can actually do we can afford." For an infrastructure-starved place like the United States, construction can generate long-run gains from higher economic activity that eventually surpass the immediate construction costs. And the experience of building big in underserved places is a means of redistribution that makes locals happy while satisfying fiscal conservatives who are normally skeptical of welfare payments.

Rather than worry about bond vigilantes, the engineering state has focused on delivering material improvements for the people. Rural folks in Guizhou have seen their material conditions of life improve immeasurably over the past few decades. The mixture of permitting free enterprise while building big infrastructure is part of the reason that the Communist Party has held on to consent of the governed.

China's policymakers have declined to be bound by some of the fundamental tenets of Wall Street investors—reduce investment, shrink assets, produce profitability—all of which emphasize efficiency. Perhaps it will trigger financial distress in the future. So far, however, building big has improved the lives of regular people, not just a narrow set of elites. This lack of emphasis on efficiency has been key to another Chinese success: Part of the reason that China dominates advanced manufacturing technologies is precisely because it tolerates lower profits while cultivating a large workforce.

CHAPTER 3

TECH POWER

I N 1980, SHENZHEN WAS best known for its oysters. For centuries, it was populated by folks who made their living from the sea: pearl fishers, salt farmers, and oyster harvesters. Villagers set cages along the coast where shifting tidal waves brought saline water, warmed by the sun, to meet cool mountain streams, producing mollusks known throughout the region for being especially succulent. That was the past. For three decades, Shenzhen's waters haven't produced oysters, their habitat flushed away by industrialization.

Shenzhen was China's greatest boomtown and, therefore, the world's. Its population soared from three hundred thousand in 1980 to seven million in 2000 and eighteen million in 2020. For many Chinese, who are intently judged on the region they're from, Shenzhen was a land of opportunity where no one was a local. One of the city's slogans, still occasionally found on billboards, reads, "You're a Shenzhen local the moment you're here." It's a poke at Beijing and Shanghai, cities where older families maintain a certain exclusivity (as they might in Paris or London).

In 1980, when Deng Xiaoping christened Shenzhen a "special economic zone," the city had little to recommend it other than its location directly abutting British-ruled Hong Kong. Deng wagered that success in Shenzhen could tear down the socialist strictures on China's economy that the rest of the leadership had been hesitating to dismantle. He lavished the city with supportive policies and penned editorials to beckon the ambitious to move there.

Answering his call were rural folks, who had never enjoyed much economic opportunity, as well as urban residents frustrated by working for rigid state enterprises. These migrants became the shock troops of China's foray into capitalism. They threw themselves into manufacturing toys, clothing, and other consumer goods in the 1980s, growing their capabilities each year. By the 2000s, Shenzhen was a major electronics hub. The workforce would become the spearhead for the greatest business endeavor of the early twentieth century: the campaign to put a smartphone into the hands of nearly everyone on the planet.

When Steve Jobs announced the iPhone in 2007, there was no more natural place than Shenzhen for mass production. It had already scaled up manufacturing of the iPod there a few years earlier. Apple decided that Shenzhen was the city to make the boldest product that Jobs had conceived.

The iPhone has become one of the rarest sorts of consumer products—both ubiquitous and coveted as a status object. It is also the crowning success of the trade relationship between two countries, in which American inspiration and marketing savvy met China's millions of workers, managed by contract manufacturers like Taiwan's Foxconn, to make state-of-the-art electronics. It wasn't easy to organize a workforce to assemble thousands of components into the most complex consumer product the world has ever known. Mastering this feat propelled Apple to become the first trillion-dollar company.

China, if anything, gained something even greater from this partnership. While the company enjoyed a surge in valuation, the country experienced a boost in national power, produced by the international collaboration needed to train hundreds of thousands of Chinese workers, every year, to build sophisticated electronics. Chinese companies subsequently leveraged this workforce to lead the world in other industries centered in Shenzhen, including electric vehicles, battery systems, and consumer drones.

As China did so, it embraced a vision of technology radically different from Silicon Valley's: the pursuit of physical and industrial technologies rather than virtual ones like social media or e-commerce platforms. In China, technology is not represented by shiny objects; rather, it is embodied by communities of engineering practice like Shenzhen, where technology lives inside the heads and in the hands of its workforce. This chapter reveals how a city climbed a technological ladder, making shirts and toys in the 1980s to making the world's most sophisticated electronics three decades later.

China, as I said in my introduction, is often messy. But in some places, it is spick-and-span. The most orderly places I've been to in the country are the manufacturing sites producing for Apple. Every worker is in place at all times. You can tell a worker's rank by their uniform: A line manager might, for example, wear green among assembly workers wearing blue. Women and men with longer hair wear hairnets. Workers are not allowed to cross into assembly lines making products for other companies. At the end of the workday, they pass through perhaps a half dozen scanners to make sure they haven't pocketed any products. A wave of people exit cafeterias or enter dormitories at designated times. Shuttles bring workers to the restaurants or karaoke spaces where they can, at last, be unregimented.

It's easy to get lost in factory zones because so many of the buildings look the same. The iPhone turbocharged factory complexes to enormous scale: Foxconn's manufacturing campus in the north of Shenzhen occupies five hundred acres. The site has factories, of course, and dormitories. It also has grocery stores, cafés, a fire brigade, a hospital, cinemas, swimming pools, and vendor-operated restaurants. The factory is the size of a city. The population peaks in early fall as production ramps up to meet demand for the Christmas season. Dormitories fill up then, with up to six men or women crammed into one room. Assembly lines operate for three eight-hour shifts a day; there is never a minute that factories aren't producing iPhones. At the peak times, three hundred thousand people work at Foxconn's Shenzhen campus, about as many as live in Pittsburgh or St. Louis. A Chinese report from 2009 estimated that the campus each day consumed forty tons of rice, twenty tons of pork, ten tons of flour, and five hundred barrels of cooking oil.

In 2020, Foxconn employed nearly a cool million workers globally. As iPhone production swung into full gear a decade ago in Shenzhen, workers might have seen someone zooming around the campus on a golf cart. That would be Terry Gou, founder of Foxconn (also known as Hon Hai Precision Industry). Gou might start the day by doing laps in the company pool and then drive his own golf cart, specially equipped with a bicycle bell, around the facility until late at night to monitor production. He is legendary in his native Taiwan for his dedication to work. Gou aggressively courted American companies like Dell and Apple to win contracts for manufacturing their products, earning their trust by guarding technical secrets and making products on time, at high quality, in massive volume.

Terry Gou also has a whimsical side. In 2019, he said that the Buddhist goddess of the sea visited him in a dream to say that he should run for president of Taiwan. In his party's primary election that year, he finished in second place.

Gou set up the officially accredited Foxconn University on the Shenzhen campus, offering twenty-five majors, most of which were engineering related. Gou surrounded himself with deputies who worked nearly as relentlessly as he did, driving Foxconn executives to the factories six days a week and then to study sessions on Sundays. In earlier years, they studied engineering principles. One former employee told me that in more recent years, political education has been more prominent, meaning that they have to study the words of China's top leader. The curriculum transitioned from "Steve Jobs thought" when Shenzhen was freewheeling a decade ago to "Xi Jinping thought" in the more disciplined present.

At the best of times, electronics assembly is overwhelmingly repetitive. Managers prize workers with daintier fingers, favoring women because they are presumed to be nimbler. When I asked factory overseers why iPhones are not made in the United States, they always bring up fingers. "Look at those meaty American hands," Taiwanese managers tell me. "How can they possibly put together something as intricate as an iPhone?"

It's hard to say what was more repetitive: studying Xi's speeches or doing electronics assembly. Both are mind numbing, but assembly work caused greater suffering. We would know far less about Foxconn if over a dozen workers in Shenzhen had not attempted suicide by jumping from factory dormitories in 2010. This tragedy forced Foxconn and Apple into crisis management mode. The press-avoidant Gou invited a few Western journalists to tour sections of the campus, which was subsequently lined with three million square meters of mesh netting woven around dormitories to prevent more deaths.

As iPhone sales started to explode, Foxconn faced a constant hunger for workers. Soon enough, it had outgrown Shenzhen. Rather than wait for migrant workers to move to Shenzhen, Gou decided to move Foxconn to the biggest suppliers of workers. Factories sprang up in China's most populous regions: Sichuan and Chongqing in the southwest, the eastern provinces around Shanghai, and the northern province of Henan. These regions remain major production sites for Apple, the biggest of which is in Henan's capital city of Zhengzhou. At peak season, Zhengzhou has the capacity to employ around 350,000 people.

Chinese officials climbed over each other to host a Foxconn facility. They salivated at the number of jobs and amount of tax revenues the company could create for their jurisdiction, which could elevate them to higher office. Local officials promised to satisfy Foxconn's extraordinary labor demands. In Chengdu, minor bureaucrats had to hit quotas on the number of workers to rustle up for factory work; those who failed might receive an order to work at assembly lines themselves. One Chengdu official who failed her quota received not just that work assignment but also cruel teasing from her more successful colleagues: "Don't leap off any buildings while you're there," someone told her.

Officials in Henan outdid themselves in hustling workers into factories. In 2016, Henan officials "borrowed" workers from state-owned coal companies to meet the iPhone production surge. In 2017, the *Financial Times* reported that up to three thousand high school students had to work on assembly lines—a few of them for eleven-hour days—and if they did not, their school withheld their graduation diplomas. They were euphemistically called "interns" who assembled iPhones for "vocational experience." In 2022, when Covid controls snarled supply chains, they recruited retired People's Liberation Army personnel to staff production lines. It was at Foxconn's Henan sites where some of the most dramatic protests against zero-Covid took place, when young men flung bricks into massed ranks of riot police.

Helen Wang (no relation) had been a Foxconn executive working in California in the early 2000s when Apple poached her to be a procurement leader. She would eventually work sourcing components for the first iPhone. In an interview, Helen told me that her first thought on receiving an assignment was often, "I need to build a city." Construction of this scale was something that Apple, Foxconn, and government officials did together. Helen told me that Shenzhen conducted leveling operations along mountains to make land suitable for production. Another former Apple engineer told me that a grassy field had turned, four months later on his next visit from Cupertino, into an industrial building with six floors getting ready to install equipment. Local officials in Shenzhen, Sichuan, and Henan not only collaborated to find labor. They also offered cheap land, extended vast tax rebates, and built roads, dormitories, and factories. The central government pitched in to help too, creating "bonded zones," which facilitated customs clearance. The state worked closely with the companies to move workers and components into factories and finished products out.

Deng Xiaoping, with the help of other reformist leaders, made Shenzhen into a hothouse of capitalism. What does capitalism need? A stock market, which Shenzhen established in 1990. What else? Belching factories with dismal labor conditions. That it had aplenty. Walmart invested deeply in that region to source goods: socks, toys, lighting, and nearly anything else that consumers wanted in a supercenter. In 2002, Walmart moved its global purchasing center from Hong Kong, a financial hub, to Shenzhen, which was closer to the factories. By that point, Shenzhen's factories had started to produce goods more sophisticated than socks. They had become proficient at developing all sorts of electronics components: small batteries, cable connectors, and display screens.

Explosive growth had costs: the oysters, for example, which could no longer live in the marine environments that the factories had spoiled. Walmart, Foxconn, and many other multinational companies have been accused of dreadful labor standards. Shenzhen built new buildings in too great haste. The government fretted about buildings that had "five lacks," that is, no design, drawings, permits, supervised construction, or official registration. The result, reported by the *Shenzhen Commercial Daily*, was that one-eighth of rural buildings completed in 1983 suffered major structural problems, sometimes including collapse.

I took frequent trips to Shenzhen when I lived in Hong Kong. You could get there via a sea ferry, which offered pleasant views, or, more conveniently, on the subway line that connects the two cities. Today, Shenzhen is one of China's most desirable places to live, gleaming with skyscrapers and malls and full of big trees. But new construction has not obliterated the city's past. Threaded between big avenues are bustling pockets of semi-preserved village structures that imbue the city with more liveliness than glass skyscrapers are able to provide. After business meetings, I would enter alleyways to find these urban villages, which have little textile workshops operating during the day and small joints serving griddles of seafood with fridges full of beer at night.

The center of Shenzhen is the Huaqiangbei mall complex. It is a giant bazaar spread across several buildings, with stalls filled not with spices or silks but wholesale electronics. Each storefront is usually made up of a brightly lit sign hanging above transparent plastic bins in which wires, specialized semiconductors, adapters, capacitors, and any electronic part imaginable can be scooped up by the armload. They buzz with the noise of activity. Clamor drifts up from the hubbub of people negotiating bulk orders, completed with the shriek that comes from ripping the packing tape that closes a box and seals the deal.

On my first visit to Huaqiangbei, I walked through the hundreds of vendors in the mall complex, when a phone case with a whale printed on it caught my eye. I decided it would be fun to carry around a reminder of *Moby-Dick*. When I went to buy the phone case, the owner was slightly taken aback that I wanted only one. "Usually, we take orders by the hundreds." It took him a moment to switch systems on his computer to accommodate my modest purchase.

Shenzhen and the surrounding cities (Guangzhou, Dongguan, Zhuhai, and a half dozen others) altogether equal the population of Germany. The area is not without its charms. Hong Kong is breathtaking with its mix of mountains and skyscrapers, while Guangzhou has marvelous temples and big villas. It's useful, however, to appreciate this region as a giant industrial complex, especially for making electronics. Drive out of downtown Shenzhen and that's easy to see. Along dusty roads, you will find factories, warehouses, and tooling shops, which are rarely beautiful and mostly drab.

It's fully appropriate to call Shenzhen the "Silicon Valley of hardware." As in the stretch from Palo Alto to San Jose, Shenzhen is full of boring office parks along highways in a beautiful natural setting. And friends would tell me that Shenzhen, as in Silicon Valley, is a great place to found a start-up. A group of people would discuss an idea over dinner, divide up the tasks, and get to work the next morning. By contrast, in Beijing, dinner will feature interminable rounds of liquor shots, reckless bluffs about connections in high places, and uncertain follow-up afterward.

It wasn't simply Apple dreaming up new ideas for its manufacturers to execute. Rather, it was a collaborative process between Cupertino and Shenzhen. "[Apple products are] not designed and sent over. That sounds like there's no interaction," Apple CEO Tim Cook once told an interviewer. The idea of having something designed in California and manufactured elsewhere "requires a kind of hand-in-glove partnership." In 2019, United Airlines made a promotional banner about how valuable Apple was to its business. United wrote that Apple booked fifty business-class seats daily from San Francisco to Shanghai, from which the airline made $35 million each year. That's over eighteen thousand business-class seats on one route.

The several dozen manufacturing sites that Apple has around the world are all meant to produce at exactly the same level of quality.

That's why Apple kept sending engineering managers from Cupertino and demanding they camp out in factories in Shenzhen or elsewhere in Asia and not come back until they had solved production issues. This demand for consistency helps to explain why the factories I visited felt so regimented: The production lines were intensely hierarchical, as extensively planned out as if they were military. It's no wonder that Foxconn's formal name is Hon Hai *Precision* Industry.

A 2012 story in the *New York Times* reported that Apple needed to hire nearly nine thousand industrial engineers in the earlier days of iPhone production. The company's analysts expected recruitment to last nine months to hire that many engineers in the United States. In China, they were able to do it in two weeks. A large pool of good labor increases the speed of design and production cycles. As Tim Cook once said, "In the US, you could have a meeting of tooling engineers and I'm not sure we could fill the room. In China, you could fill multiple football fields."

Apple and Foxconn found an advantage in Shenzhen beyond workers who could meet their quality standards: The dense network of factories also offered flexibility on manufacturing techniques. One of the former Apple engineers I spoke to pointed out that any feature changes create unpredictable demands. Each year, Apple might have a pretty good sense of where most valuable iPhone components are coming from (the camera module, for example, from Sony; the memory from Samsung; its chips made by TSMC), but there are constant surprises further down the supply chain. "There are always new components or processes that a new design requires, like a certain type of adhesive or a screw of a slightly different size."

Therefore, Apple constantly had to scramble to find suppliers on short notice. "Almost always," the engineer continued, "we found someone in Shenzhen by asking a guy who knows a guy whose cousin might be able to produce a few hundred thousand new screws."

Virtually everything one needs to produce any electronic product can be found in a short drive around Shenzhen. Proximity creates efficiency. When it's time to do stuff, a company can collapse coordination that usually takes weeks into a business meeting lasting hours by convening all the relevant suppliers in one room the next morning. And if something goes wrong, there are a lot of friendly neighboring factories to call. "If you have a gas leak," an American hardware entrepreneur told me, "you can go borrow a neighbor's kit and give it back the next day."

Workers in Shenzhen gained skills by assembling smartphones, music players, and other electronics. It didn't take long for some engineers and line managers to rummage around the plastic bins of Huaqiangbei, wondering what they could do with these parts. These components were getting better every year, part of a trend that Chris Anderson, former editor of *Wired*, called "the peace dividends of the smartphone wars." The hundreds of billions of dollars invested in the smartphone supply chain have caused the cost of electronic components—cameras, sensors, batteries, modems—to plummet. That's why we're able to carry around sensors in our pockets that used to be available to only a select few military powers.

Many companies have grown around this peace dividend. Indeed, Shenzhen is the headquarters of many of China's most dynamic companies, including BYD, the world's largest electric vehicle maker; DJI, the world's largest consumer drone maker; and Huawei, the beleaguered company that is the world's largest telecommunications equipment maker. Electric vehicles are full of the electronic components borrowed from smartphones; the consumer drone is roughly a reassembly of a smartphone camera and sensor with propellers for flight.

The magic of Shenzhen is the combination of the world's most creative hardware engineers sitting in a sea of components that improve every year amid a labor force of millions who know how to put together electronics. This buzzing ecosystem has produced many other products that follow in Apple's wake, like hoverboards, electric scooters, virtual reality headsets, and who knows what's next?

• • •

WHEN I MOVED to China in 2017 to cover technology, it was still common to hear Americans say that Chinese companies couldn't innovate. China could only copy and steal, they said. Some folks in Silicon Valley knew that there were cool things cooking in Shenzhen, but the broader attitude among Americans was condescension.

When I left China in 2023, the tenor of American views had shifted. Fewer people were saying China hasn't developed any important technologies, since it has become a major producer of electric vehicles and clean technologies. Alarm has crowded out the dismissiveness, as China's surveillance capabilities are menacing US national security while its manufacturing capacity is threatening to engulf Western firms.

We are still not appreciating the communities of engineering practice like Shenzhen, and at no point has there been real curiosity about how China's technological capabilities have developed.

The iPhone embodies China's steady technological ascension. In 2007, Apple imported nearly all of the high-valued components— display screen glass from the United States, camera modules from Japan, memory chips from South Korea, sensors from Germany—to Shenzhen. China's contribution consisted mostly of the labor involved in assembling foreign products, which was around 4 percent of the phone's final value. One former Apple executive told me that the iPhone supply chain grew more "red" over the next decade as it incorporated domestically produced components—meaning that it incorporated more Chinese components. By the time that the iPhone X was released in 2017, Chinese firms were making acoustic parts, charging modules, and battery packs. According to a teardown analysis, China's contribution to the iPhone X reached around 25 percent of the final value of the phone.

In the 2010s, China produced the digital platforms that Americans have associated with real technological innovation. In 2017, tech giants like Alibaba and Tencent brawled with each other, as well as with up-and-coming firms like ByteDance, for the billion Chinese users who were getting online. E-commerce companies like Alibaba held ludicrously fun sale bonanzas, hiring Taylor Swift to perform a concert in Shanghai to drive a buying frenzy. Chinese consumers were some of the most eager adopters of online retail in the world; since they live in dense cities with superb logistics networks, platforms were able to deliver goods rapidly. People skipped several steps in Western habits, dispensing with personal computers, email, and credit cards so that they could manage their lives on their smartphones, especially through Tencent's WeChat app. In 2017, TikTok was gaining traction, and China looked like it might be strong on AI and maybe dominate Bitcoin too, given that most of the world's mining servers were there.

A few years later Xi Jinping kneecapped most of China's digital platforms. Xi prefers his industry heavy and his output hard. He scorned the virtual economy, denouncing the "barbaric growth" of capital and focusing instead on industrial developments. That meant throwing everything into manufacturing. Though it remains several steps behind the West in a few critical industries, especially semiconductors and aviation, Chinese manufacturing has caught up in most other fields.

China leads the world in deploying ultrahigh-voltage transmission lines, high-speed rail, and 5G networks. Chinese manufacturers make machine tools—die-casting machines, steel presses, robotic arms—that approach German and Japanese levels of quality. They've muscled out most other Asian competitors on consumer electronics. Phone makers like Huawei, Oppo, Vivo, and Xiaomi tapped into the worker and component ecosystem that Apple helped to build. In 2025, the world's largest phone makers are Apple, Samsung, and a half dozen Chinese firms that concentrate on sales to developing countries.

Chinese brands are not only making many of the lowest-end consumer goods (the junk found on e-commerce apps) but also higher-end kitchen products and audio equipment. It's fair to say, however, that although Chinese workers make so much stuff, few Chinese companies have established striking global brands. They're far behind Japanese companies, which, starting in the 1970s, created whole new categories of products, like music players, game consoles, digital cameras, and pocket calculators, that excited global consumers. For the most part, Chinese successes involve making good products cheaply. But I think it's likely that they will be known for great products too. Branding tends to follow good quality, and I expect Chinese brands to be well regarded over the next decade, just as the perception of "Made in Japan" flipped from shoddy to valuable.

China's clearest industrial success involves clean technology, or the renewable power equipment that we need to decarbonize our economies. In 2025, Chinese firms dominate every segment of the solar value chain, make most of the large-capacity batteries that power electric vehicles, and have commanding positions in wind turbines and hydrogen electrolyzers.

China remains weak in several industries, however. The leadership is sore that the country remains dependent on the West for aircraft engines and semiconductor technologies. And though China's biotech industry is big, Chinese pharmaceuticals haven't yet produced a blockbuster new drug or vaccine. Not often do its universities generate groundbreaking new papers that force American scientists to sit up and pay attention.

The fact is that China remains fairly weak at producing scientific advancements. Whereas Japanese researchers have earned more than twenty Nobel Prizes in the sciences, only one has ever been awarded to a Chinese national. Now the state is dedicating enormous resources to pursuing better science. In 2019, China became the first country to land a rover on the far side of the moon; a year later, Chinese scientists achieved quantum-encrypted communication by satellite. Its space agency has announced that it will land people on the moon by 2030. That's hardly outdoing the United States in space, which landed astronauts on the moon six decades before China's target. But it is a sign that China is steadily investing in scientific capabilities that give it the power to achieve increasingly difficult tasks.

It's another of the ways that the United States and China are inversions of each other. Americans expect innovations from scientists working at NASA, in universities, or in research labs. They celebrate the moment of invention: the first solar cell, the first personal computer, first in flight. In China, on the other hand, tech innovation emerges from the factory floor, when a new product is scaled up into mass production. At the heart of China's ascendancy in advanced technology is its spectacular capacity for learning by doing and consistently improving things.

. . .

WHEN WE TALK ABOUT technology, we should really distinguish between three things. First, technology means tools. These are the pots, pans, knives, and ovens required to prepare a dish. Second, technology means explicit instruction. These are the recipes, the blueprints, the patents that can be written down. Third and most important, technology is process knowledge. That is the proficiency gained from practical experience, which isn't easily communicated. Ask someone who has never cooked before to do something as simple as fry an egg. Give him a beautiful kitchen and the most exquisitely detailed recipe, and he might still make a mess.

We can see how China values process knowledge through its approach to architecture too. That reveals something deeper and more interesting about its culture. One of my favorite books about China is a collection of essays called *The Hall of Uselessness* by the Belgian sinologist Simon Leys. In one of these essays, "The Chinese Attitude Towards the Past," Leys considers the construction techniques of Chinese builders.

Builders everywhere have attempted to overcome the erosion of time. Ancient Egypt and medieval Europe built great pyramids and cathedrals out of stone. The approach in China, as Leys points out, is for builders to yield to the onrush of time by using eminently perishable, and indeed fragile, materials. By building temples out of wood with paneling sometimes made of paper, Chinese architecture has built-in obsolescence, demanding frequent renewal. "Eternity should not inhabit the building," Leys writes. "It should inhabit the builder." Rather than using the strongest materials, Chinese builders have embraced transience to ensure the eternity of spiritual designs.

The shining exemplar of this idea is found not in China but at the Ise Grand Shrine (or Ise Jingu) in Japan. Ise Jingu is the holiest shrine in Japan's Shinto faith. Since it was first erected in 690 AD, craftspeople have completely rebuilt its sacred temples—made of wood and hay— every twenty years. In 2033, the temple will be rebuilt for its sixty-third reconsecration. Ise Jingu's halls are made of Japanese cypress timbers that support a raised floor and are covered by a thatched roof of dried silvergrass. These structures use techniques from the seventh century: no nails, only dowels and wood joints. Though wood joinery is a complex craft, the rest of the construction is simple.

Why does this ritual persist? In part, it has to do with the Shinto faith in spiritual renewal. And it is also because these shrines are built in the style of rice warehouses, dedicated as they are to the god of agriculture, which rot every few decades. It is also about the preservation of craft knowledge. Twenty years is the length of a generation, and the caretakers of the Ise Jingu have attempted to ensure that knowledge about how to rebuild this shrine can be passed on to descendants. Junko Edahiro, an environmental writer who witnessed the sixty-second rebuilding, heard an elderly fellow say to younger folks, "I will leave these duties to you next time."

Edahiro wrote a piece entitled "Rebuilding Every 20 Years Renders Sanctuaries Eternal." Shrine staff make plans measured in centuries: They have a two-hundred-year road map to plant enough cypress trees to make the nearby shrine forest self-sufficient, rather than having to ship timber in from other parts of Japan. Their planning and the ritual make me wonder how much process knowledge the West has given up. When a fire broke out on the roof of Notre Dame de Paris in 2019, it revealed how little knowledge about cathedral construction is left in the world. I would bet that Ise Jingu, built out of wood, will endure longer than the great pyramids and cathedrals made of stone.

Embracing process knowledge means looking to people to embody eternity rather than to grand monuments. Furthermore, instead of viewing "technology" as a series of cool objects, we should look at it as a living practice. That is closer to the approach used in China and Japan.

If Japanese craftspeople have put in this much work to retain knowledge of a seventh-century temple, how are we supposed to maintain the vast technological civilization we've built? This wooden structure is so much simpler than a modern auto plant, to say nothing of a semiconductor fab. Can we moderns preserve manufacturing knowledge without enacting the rituals of craftspeople?

The answer, perhaps, is that we can't. It's not just Boeing and Intel that have lost their way. In the time it took to do one rebuild of the Ise Jingu, the US government forgot something only as important as nuclear weapon material. The National Nuclear Security Administration found that it could no longer produce "Fogbank," a classified material used to detonate the bomb, because it hadn't kept good records of the production process and everyone who knew how to produce it had retired. The NNSA then spent $69 million to relearn how to produce this material.

It's rare for blueprints to encode enough information to be technologically valuable. Imagine if we were able to send the most detailed instructions for building any modern technology back to the past. The lead chariot engineer of a Roman caesar would get nowhere with the most detailed manual and finely drawn blueprints on how to produce a Model T. Nor would many of us in the present be able to do much if we got our hands on the instructions for producing an Intel processor or ASML lithography machine. I am not proud to have struggled with putting together a footrest from IKEA.

Process knowledge is hard to measure because it exists mostly in people's heads and the pattern of their relationships to other technical workers. We tend to refer to these intangibles as know-how, institutional memory, or tacit knowledge. They are embodied by an experienced workforce like Shenzhen's. There, someone might work at an iPhone plant one year, for a rival phone maker the next, and then start a drone company. If an engineer in Shenzhen has an idea for a new product, it's easy to tap into an eager network of investors. Shenzhen is a community of engineering practice where factory owners, skilled engineers, entrepreneurs, investors, and researchers mix with the world's most experienced workforce at producing high-end electronics.

Silicon Valley used to be like this too, but now it lacks a critical link in the chain—the manufacturing workforce. The value of these communities of engineering practice is greater than any single company or engineer. Rather, they have to be understood as ecosystems of technology.

The American imagination has been too focused on the creation of tooling and blueprints. Andy Grove, the legendary former CEO of Intel, said it best in 2010: that the United States needs to focus less on "the mythical moment of creation" and more on the "scaling up" of products. Grove saw Silicon Valley transition from doing both invention and production to specializing only in the former. And he understood quite well that technology ecosystems would rust if the research and development no longer had a learning loop from the production process.

The United States does want to re-create Shenzhen's success. But it has had, at best, a surface-level understanding of its success. Silvia Lindtner, a professor at the University of Michigan and my wife, has spent more than a decade studying Shenzhen's technology ecosystems. In 2015, the Austrian government asked her how to create a Shenzhen in the Alps; in 2016, the White House invited her to present on how the United States might learn from the success of Shenzhen. She has felt, as I do, that these agencies misunderstood the point of Shenzhen. They were still more interested in individual inventors rather than understanding it as a community of engineering practice. The obsession with invention has clouded Silicon Valley's ability to appreciate China's actual strength. Rather than seeing tools and blueprints as the ultimate ends of technological progress, I believe we should view them as milestones in the training of better scientists and manufacturers. Viewing technology as people and process knowledge isn't only more accurate; it also empowers our sense of agency to control the technologies we are producing.

· · ·

VIEWING TECHNOLOGY AS PEOPLE also helps us understand why economic relations between the United States and China have broken down. Through the 1990s and especially after 2001 (when China acceded to the World Trade Organization), American companies were busy moving manufacturing work to China. Apple's collaboration in Shenzhen helped transform the city into the world's most innovative hub for electronics production. But this win for Apple's shareholders has been a loss for American power.

US manufacturing employment peaked in 1980 at nineteen million workers. In 2000, it still had seventeen million. Then it collapsed over the next decade, in part due to China, in part due to technology changes, and especially after the global financial crisis, when the workforce fell to just eleven million in 2010. In 2025, the United States has around thirteen million manufacturing workers.

At times, American elites have been strangely good humored about the departure of manufacturing jobs. In 1993, the chief economic adviser to George H. W. Bush, Michael Boskin, quipped, "Computer chips, potato chips, what's the difference?" It became part of the elite consensus that the United States could lose manufacturing. This consensus portrayed union bosses, as well as the handful of heterodox economists, as sentimentalists for resisting offshoring. Neither the Clinton nor George W. Bush administration restrained American firms from moving manufacturing operations to China. Now, it's more obvious that the departure of manufacturing has created economic and political ruination for the United States. We are still only beginning to understand how much it set the country back technologically.

Many of the United States' most storied companies have been ailing. Detroit's automakers, having limped along for decades, are now stumbling through the transition to electric vehicles. US Steel, General Electric, and IBM are shadows of their past selves. Intel, mired in cycles of blown product timelines and layoffs, went from a semiconductor trailblazer to a clear laggard behind Taiwan's TSMC. After two of Boeing's 737 MAX jets crashed in 2017, the company promised strenuous efforts to guarantee the safety of its aircraft. Then a door blew off midair in 2024. Boeing, like Intel, is constantly delaying the launch of long-planned products.

Even the military-industrial complex looks challenged. The United States spends nearly $1 trillion a year on defense, about as much as the next ten countries combined. The return on this investment is not clear. In the aftermath of Russia's invasion, Ukraine blew through several years' worth of American munitions stockpiles in a matter of months, and American factories have struggled to scale up production. Fighter jets have faced enormous delays and cost overruns. The US Navy has reported that every single class of its ships and submarines is one to three years behind schedule.

American manufacturers aren't all languishing. Tesla is America's great hope in automaking. There remain many leaders across semiconductor production equipment makers, medical devices, and agricultural equipment. The great success of the US manufacturing sector over the past several years was the production of mRNA vaccines, which have saved lives around the world. But the triumphs in medicine and pharmaceuticals were not matched by the broader set of American manufacturers, who failed to produce basics like masks and cotton swabs.

The US manufacturing base has, with some exceptions, rusted from top to bottom. Why have so many manufacturers crumbled? Partly, I think, we can examine the culture of financial investors. Wall Street has been far keener to invest in capital-light businesses: digital platforms like social media and search engines or chip companies that focus on design rather than cumbersome fabrication facilities. If it weren't for Tesla (which makes many of its cars in Shanghai), the United States would be even further behind China in electric vehicles. And Tesla's survival was a close-run thing. In 2018, Elon Musk said that Tesla was on the verge of bankruptcy as it tried to ramp up the production of the ultimately successful Model 3. It was a time he called "excruciating." In retrospect, it is an indictment of the American financial system that fundraising for a manufacturing leader had to be this difficult. Financialization also intersects with corporate consolidation. One prominent line of argument regarding General Electric was that the company was taken over by finance. That applies in greater force against Boeing. Once run by engineers obsessed with safety and quality, its leadership shifted to executives more focused on delivering shareholder value than good planes.

Mostly, though, I think the problem lies with American policy-makers and executives who fail to grasp the importance of process knowledge.

American manufacturers spent the better part of three decades unwinding its stock of process knowledge when it opened so many factories in China. Every US factory closure represents a likely permanent loss of production skill and knowledge. Line workers, machinists, and product designers are thrown out of work; then their suppliers and technical advisers struggle as well. Entire American communities of engineering practice have dissolved, leaving behind a region known as the Rust Belt. Some mayors and governors tried to stem this receding tide. But they were continuously scorned by economists and executives, who sought low-wage production in the name of globalization. Still today, many American economists doubt there is anything special about manufacturing and put their faith in the inevitable march to a service economy.

Low-wage ecosystems like Shenzhen became a giant magnet for US process knowledge. Beijing made a deliberate decision not to be like Japan, which kept its market limited to American companies; rather, China mostly welcomed foreign manufacturers to train its workers. It is some sign of China's economic openness that so much of its exports are driven by Apple and Tesla, while Japanese exports have been driven almost entirely by its own companies. After it built up a critical mass of process knowledge, however, Shenzhen became as much an innovator of new electronics as an implementer of American ideas.

It's not clear to me that it was part of Beijing's grand strategy to rely on American companies to become a manufacturing leader. But in some cases, the state understood that's what they were doing. Beijing did something unprecedented for Tesla in 2018: It allowed the company to fully own its plant in Shanghai. Previously, any automaker that wanted to produce in China had to partner with a domestic company. So Japanese, German, and American companies dutifully partnered with state-owned enterprises in order to access the enormous market. The state had hoped that these domestic companies would learn from the likes of Toyota and Mercedes-Benz and match their quality. In reality, Chinese automakers were sluggish from their research dependence on their foreign friends.

Tesla's presence jolted China's electric vehicle market. China's business community began using the term "catfishing" for what Tesla was doing in China. The idea was that introducing a powerful new creature into the domestic environment would make Chinese firms swim faster. That's exactly what they did to raise their game. When Tesla vehicles started rolling out of the Shanghai Gigafactory in 2019, BYD saw its sales decline by 11 percent, while profits fell by 42 percent. But Tesla would eventually do the whole market a favor. As in the United States, the company's audacious branding stimulated consumers to think of electric vehicles as more than high-powered golf carts. And Tesla made investments in China's tooling ecosystem that other automakers exploited to produce better cars. BYD benefited as well, reporting record profits in 2023 and becoming the world's largest electric vehicle maker. And even the Communist Party's main newspaper praised how Tesla produced the "catfish effect" for Chinese firms.

As Grace Wang, founder of Shenzhen-based Luxshare (one of Apple's new contract manufacturers), poetically expressed, "Flying with phoenixes will nurture outstanding birds." It is another lesson that capitalist Shenzhen has taught the Communist Party: Market competition tends to lower prices and raise quality.

Apple and Tesla have made a huge effort to train its Chinese workers to manufacture their products—and earned fabulous sums of money by doing so. These stories are replicated in varying degrees across China's other communities of engineering practice, production hubs for shoes and garments in the eastern city of Wenzhou, medical equipment in Wuxi and Suzhou, and, most wonderfully of all, guitars in the mountains of Guizhou's Zheng'an County. Overall, China's manufacturing workforce employs more than a hundred million people, around eight times that of the United States. That is a big stock of people who are fueling the creation of new process knowledge.

A focus on manufacturing gives China another advantage in technological competition with the United States. It can simply wait for American scientists to do the fundamental research before Chinese companies take over the production. That is, in essence, what happened with the solar industry. Bell Labs invented the first solar cell, and German companies produced solar production equipment. Beijing's designation of solar as a "strategic emerging industry" invited Chinese companies to rush into this industry. Chinese companies bought German equipment and competed fiercely to make the most efficient solar cells. By the mid-2010s, Chinese companies figured out how to make all the German tools, as well as the entirety of the solar value chain. The plunge in solar power costs over the last decade has been driven less by breakthroughs in science—which is the United States' strong suit—than by efficient production, which is China's strength. The beneficiaries are not only the climate but also China's national power.

Science matters of course. China remains weak in chips and aviation in part because these are much more scientifically complex industries than solar. Not every technology improves through iterative adjustments to manufacturing processes, but a great deal can follow its logic. When lots of companies are doing similar things, in a brutally competitive environment where profit margins are small, they establish communities of engineering practice like Shenzhen. These factories will never be as glamorous as the desirable branding represented by Apple or Tesla. Every day, millions of workers go to factories to build up technological process knowledge. That is the basis of China's tech power.

• • •

CHINA HAS BECOME A tech superpower by exalting process knowledge and the communities of engineering practice that keep it alive. Holding on to process knowledge helps us resist bad ideas about China's rise. The Communist Party would love to claim that China's technology sector developed the way it has through wise planning from Beijing. And the American government also overstates the importance of the Chinese government through its accusations of cheating (including with unfair subsidies) or stealing (especially through cybertheft).

The results of the Chinese government's unceasing interventions in the economy are at best ambiguous. Economic studies have shown that the recipients of Chinese subsidies have, on average, lower productivity growth. Xi's aggressive promotion of industry has triggered trade wars with not just the United States but also many developing countries as well. China's tech successes are no convincing demonstration that a wise state can plan the future. When the state shoves its weight around—forcing foreign companies to hand over technology, showering a favored sector with subsidies, injuring a firm while elevating another—it is often far from being helpful. The forced technology transfer agreements meant to prop up China's state-owned automakers instead robbed their need to invest in their own innovative capacities. China's automotive successes come from companies like privately owned BYD, which had no foreign partners, after the entrance of wholly owned Tesla forced the company to raise its game.

American administrations have complained about a host of China's trade practices: forced technology transfers; currency manipulation that keeps exports cheap; subsidies and generous credit terms for local firms, sometimes funding their expansion overseas; and, worst of all, unauthorized cyber intrusions, or the state-directed hacking to steal US trade secrets. Overall, they create an environment for foreign businesses that is often unfair and sometimes baffling.

In response, the first Trump administration launched its trade war. But it didn't just levy tariffs on Chinese goods. It expanded and deployed novel technology controls meant to cripple Chinese firms. While I was covering the impacts of Trump's tech war from Beijing, I remember often waking up to wonder which Chinese company he might be tweeting about. China's tech leaders found themselves designated to sanctions lists maintained by obscure US government agencies that few US officials had even heard of. Once they're on a list, which blocks American funding or technologies, it's hard to get off.

. I thought that the US government was right to push back against China's mercantilist trade practices. But I also thought that it was doing so in mostly ineffective ways under Trump's chaotic direction. In particular, I was skeptical of the security-based view of the Trump administration (as well as the successive Biden administration): that the United States still controls a lot of technological chokepoints, if only the government weren't asleep at the wheel while China stole its way to primacy; and that if the US government stepped hard on export controls, it would be able to recover technology primacy from a country that cannot match American ingenuity.

The Trump administration certainly throttled Chinese companies. But it did so by making American companies (especially those selling semiconductors) unreliable vendors. Previously, Chinese companies bought the best components on the market, which were often American, because they wanted to sell a globally competitive smartphone or drone. When they couldn't buy American, it lit a fire under Chinese companies to try domestic vendors that they would never have previously given the time of day.

When I worked in Silicon Valley, people liked to say that knowledge travels at the speed of beer. Engineers like to talk to each other to solve technical problems, which is how knowledge diffuses. They are poached by rival firms or sometimes rival countries. Over the longer run, it's difficult for countries to monopolize their dominance over any technology. If such a thing were possible, then the United States would still be behind the United Kingdom or Germany, which were much greater scientific innovators.

The US government has indulged a preening self-regard concerning how much technological power its country still wields. American companies have spent two decades building communities of engineering practice in China, made up of people who roll up their sleeves to figure out how to overcome their daily bottlenecks. It wasn't going to be easy to stop their progress; if anything, American policies risked accelerating it. So far, Chinese companies have managed to innovate around most technological restraints; rather than face precipitous collapse, as US policymakers predicted, some have even managed to keep growing at a healthy clip.

· · ·

FOREIGN COMPANIES SEEDED the initial growth of zones like Shenzhen two decades ago. Now, the relationship between the United States and China has soured. Does that mean that communities of engineering practice like Shenzhen will wither? Yes, but not for a long while.

The process of extricating manufacturing production from China will be prolonged and halting. International companies continue to tell me that they are still reluctant to completely pull up their roots from what remains an extraordinary production hub and a very big market. Apple is making immense efforts to cultivate production sites in Vietnam and India. But it is going to be gradual, since the infrastructure and labor in these countries will take a while to catch up. According to Apple's most recent supplier report (released in 2023), 156 of its top 200 suppliers have manufacturing sites in China. Seventy-two of them are in Shenzhen's province of Guangdong, which is as many as there are in the United States, Vietnam, and India combined.

Meanwhile, Xi Jinping is insistent about holding on to manufacturing. China's Communist Party might be the most technology-obsessed institution in the world. The engineering state is intent on achieving tech primacy before multinationals pull away.

On a 2023 inspection tour through Jiangsu province (like Guangdong, a manufacturing powerhouse), Xi said, "The real economy is the foundation of a country's economy, the fundamental source of wealth creation, and an important pillar of national strength." It is the basis, he continued, of "human production, life, and development." He has repeatedly said that China needs to prioritize the real economy, which means the world of manufactured products, rather than the virtual or financial economy, sometimes referred to in state media as the "fictitious" economy. State-affiliated researchers commonly denounce financialization with the hollowing out of manufacturing in the same breath.

Xi isn't just ambitious about manufacturing. A better word to describe his views might be "completionist." Andrew Batson, research director at Gavekal Dragonomics, came upon a 2024 boast from the minister of industry and information technology that China has a "comprehensive" industrial chain, since it produces something in each of the 419 industrial product categories maintained by the United Nations to classify industrial production. It's a very Chinese sort of boast.

Batson has furthermore detected a shift in Xi's rhetoric on manufacturing. Previous Chinese leaders have talked about the importance of upgrading industry, which sometimes means limiting investment into labor-intensive or highly polluting sectors that China no longer needs. Xi has declared that China targets completionism, which means that not even "low-end industries" should move out of China. Rather than follow economic logic, in which production gravitates toward countries with lower labor costs—which the United States and other high-income countries have more or less accepted—Xi does not want industry to keep shifting around.

So the Fourteenth Five-Year Plan released in 2021 demands that the manufacturing share of the economy stay constant. Manufacturing already accounts for 28 percent of China's GDP, which is much higher than Germany's 21 percent and Japan's 20 percent, to say nothing of deindustrialized economies like the United States and the United Kingdom (both around 10 percent). Xi has repeatedly stated that he's not interested in abandoning manufacturing for services. In authoritative speeches, Xi cited "certain Western countries" that forsook the real economy for the fictitious economy. No points for guessing which Western countries these might be. And Xi has declared that "the real economy is the basis of everything . . . so we must never deindustrialize."

That is what the engineering state is about. It likes to build not just public works but also manufacturing capacity. The engineering state resists economists as easily as lawyers. Economists may cite David Ricardo's theory of comparative advantage as a reason to permit production to move away. The engineering state declines, aghast at losing manufacturing because it's somehow cooler to be in services.

So far, China hasn't felt the economic pressure to abandon low-end manufacturing (clothing, footwear, and so on), in part because there are still a lot of poor Chinese provinces like Guizhou that have cheap labor. That trend might not hold given escalating tariffs. But if Xi is successful, it means that other developing countries (in Asia, Africa, and around the world) will be unable to climb the industrial ladder that China reigns over. Developed countries have reason to be alarmed as well. Since China is so large, it has the financial firepower to target any industry it wants for technological leadership. Small countries have had to pick their battles, as Denmark did in the wind industry and South Korea did with memory chips.

China wants to have it all.

China's political leadership has long cherished its hatred of Western domination and nurtured its fantasy that the country could have succeeded if only it had science, technology, and industrial production. Every Chinese leader since the Qing emperors who lost the Opium Wars has felt aggrieved about falling behind in technology. Maintaining an industrial base is the best guarantee that China won't lose again. This thread runs through China's modern leaders, from Nationalist Sun Yat-sen, his protégé Chiang Kai-shek, and then the Communist rulers too. Deng Xiaoping launched his great project to unshackle China from socialism by appealing to the Four Modernizations: agriculture, industry, defense, and science and technology. In recent years, Xi Jinping has issued increasingly urgent calls to make China advanced and self-sufficient in technology, though often in bland mouthfuls like the "innovation-driven development strategy" or the Marx-inspired "new productive forces."

An obsession with technology has spawned what is perhaps the most interesting online movement in China. In the heavily censored realm of the Chinese internet, where no group is allowed to be very organized, one set of intellectuals has made themselves heard. They are loosely affiliated writers who refer to themselves as the Industrial Party. Their views are simple to summarize: that nation-states ruthlessly compete with each other; that science and technology are the decisive forces in this Darwinian competition; and that therefore the state must be organized around the pursuit of science and technology. They patriotically view the Communist Party as the world's most capable political organization for this pursuit.

I've spent months reading some of the foundational texts around the Industrial Party. A few of these works have English translations, but most are left untouched, with much of the writing consisting of screeds on online bulletins. They tend to carry a combative tone that scorns liberals, advocates of democracy in China, and, sometimes, leftists who yearn for Mao. They set themselves against those guilty of romanticism, which they label as the Sentimental Party.

The stalwarts of the Industrial Party have diverse backgrounds. The eldest member, Wang Xiaodong, introduced the party name in an online essay in 2011. Wang had trained as an economist and found his calling as a fierce nationalist: Since the 1990s, he has written scathing books calling for China not to slavishly follow Western (and mostly American) values, culminating in a bestseller, *China Is Unhappy*, which issued a blunt call to take a more confrontational approach with the American-led order.

Zhong Qing trained as an electrical engineer in Japan and developed his views by establishing an early presence on China's online bulletin boards. His 2005 book *Wash Dishes or Study?* called for full technocratic control over the economy in order to pursue science and technology. That meant forgoing low-end manufacturing to pursue a crash program building fighter jets and semiconductors. The most active contributor to the Industrial Party ideas over the past few years is a pseudonymous writer named Shenzhen Ningnanshan, who describes himself as a middle-class person based in Shenzhen, who might be working with a state-affiliated think tank. Shenzhen Ningnanshan's articles are fully in line with the Chinese state's orthodoxy, advocating for a gradualist approach to science and technology investment, with a focus on semiconductors in order to break the US stranglehold on this technology. That makes him more of a political moderate in the Industrial Party.

Perhaps the most interesting way that the Industrial Party's ideas have been propagated is through an online novel, *The Morning Star of Lingao*, which has been serialized by a group of authors since 2009. It is an alternate-history project that imagines that five hundred people from contemporary China traveled back in time to Lingao County in Hainan (the tropical island that is China's southernmost province) in the year 1628. Their goal? To trigger an industrial revolution in the Ming dynasty. Ma Qianzu is a writer involved in the early creation of this series and is one of the more interesting personalities on the Chinese internet. Ma propelled the Industrial Party toward a breakout moment in 2011: After China's deadliest train collision, he forcefully advocated that the state should press forward with its development of the high-speed rail program (which it did). Ma is also a thinker with an independent streak. In recent years, he has exposed wasteful government spending and has been critical of Russia's invasion of Ukraine. These unusual positions have sometimes landed him in the censors' crosshairs.

None of these writers would proclaim himself a card-carrying member of the Industrial Party. They are loosely connected bloggers only sometimes in conversation with each other. Ma Qianzu has rejected the label of Industrial Party, and Wang Xiaodong has renounced some of his earlier nationalism. In recent years, he has said that China is not yet ready to sever ties with the West. A few of these writers work in academia and think tanks, which suggests direct ties to the policymakers; some of their views are reprinted in China's state media. A few, interestingly, have studied in Japan, calling for China to imitate its wartime tormentor. Many are military nerds, who know by heart the specs of different leading fighter jets. For them, there is no problem heavy industry cannot solve.

I wonder, when I read these works, whether the Industrial Party is a modern name for an old idea. These writers have a futurist bent, they denounce liberal niceties, and they demand total mobilization of the economy to pursue science and technology.

Are they simply reinventing fascism? The Industrial Party wants to depoliticize society to enable rule by technocrats, who would wield the propaganda organs to motivate people to pursue science and manufacturing. They are a heavily male group that mocks pluralism. They are not advocating conquest, but they do pine for a future in which China is stronger than any other nation. The Industrial Party tends not to cite a broad range of thinkers, only forceful leaders like Mao or Stalin who repelled invaders and established an industrial base. It is a worship of strength through technology.

The one work that much of the Industrial Party has rallied around is the science fiction trilogy by Liu Cixin. *The Three-Body Problem* is one of China's most successful cultural exports in the past decades, earning praise from American readers as well as a big-budget Netflix adaptation. I have been deeply drawn in by the trilogy myself. Its premise is that a victim of Mao's Cultural Revolution grew so disgusted with humanity that she invited extraterrestrials to conquer human civilization; when her action is discovered, governments have a few decades to prepare for the invasion.

Liu's story spans not only galaxies but also eighteen million years. He created startling images: a silver probe resembling the shape of a water droplet that destroys most of Earth's spacefleet; a particle the size of a proton that contains a whole world; a starry sky that flickers for a single observer. Major characters struggle with strategic questions involving deduction and deception, and a wrong move could be fatal not to the individual but to humanity at large.

The morality of the *Three-Body* trilogy is, meanwhile, animated by the bleakest of worldviews. On one level, the trilogy is a celebration of humanity's ingenuity in an existential struggle. To defeat the alien threat, Liu depicts humanity's total subordination to technocratic authorities. Scientists and engineers are the ultimate decisionmakers, leaving no room for humanists, the faint of heart, or sentimentalists. Governments are made to submit to the will of select geniuses who do not hesitate to sacrifice millions. The prevailing idea in Liu's trilogy is that the only hard truth is survival, where opposing civilizations resemble "blood-drenched pyramids lit by insidious fires seen through dark forests." Again and again, Liu resolves the plot in favor of the party that is willing to be the most brutal in its will to survive.

· · ·

IT'S EASY TO SEE why Industrial Party enthusiasts have elevated Liu's work to the top of its canon. It may as well be, in addition, a guide to the ideology of the engineering state.

China took up a lot of the dirty industries that the United States was happy to get rid of. In some cases, literally: Rare earth metals are not really rare. Processing them, however, demands enormous amounts of energy and water while spewing carcinogens into the atmosphere. Few parts of the Western world have the stomach for processing rare earth metals, which is why China controls this supply chain.

Most forms of low-end manufacturing aren't as bad as that, but the United States was just as willing to let them go, with little understanding of how much it would hurt the country. It's hard, I admit, to draw a straight line between the loss of, for example, television manufacturing in the United States through the 1980s to the stumbles by Boeing and Intel over the past decade. But if we think about technology ecosystems as communities of engineering practice, it makes sense that factory closures accelerated as process knowledge dissolved, prompting production problems and more job losses. And it also makes sense that Chinese workers went from merely assembling iPhones to producing some of their most valuable components as well. As one country lost its process knowledge, the other gained whole industries.

The United States has changed its mind on policy: It wants manufacturing jobs back. But how to achieve that is terribly unclear. Tariffs under Trump and subsidies under Biden haven't decisively moved the needle. Indeed, China's goods exports to the United States hit a near record in 2022, the same level as in 2018, when the Trump administration initiated tariffs on China.

How can the United States do better? As a starting point, it could develop a better understanding of how China has grown into a technology superpower. If members of Congress continue to resort to the laziest explanations ("they're just stealing all our IP"), then the United States will never grasp the importance of building up process knowledge. And it will fail to gain urgency to fix its technological deficiencies.

At the same time, Americans should develop a bit more humility about their own technological capabilities. The sooner that the United States treats China as a peer worth studying, the sooner it can develop a new playbook for success. Chinese companies are currently beating the rest of the world in the production of electric vehicle batteries. So why not allow a few of them to build factories, as they are trying to do, in states like Michigan, and force them to give up their technology? The US government could force Chinese battery makers to transfer intellectual property in exchange for accessing the giant US market for cars.

And which types of technologies the United States should pursue is also worth meditating on. Should it really go all in on artificial intelligence, cryptocurrencies, and other things that the Communist Party mocks as the fictitious economy? Or should it pursue the sorts of heavy industry that have long fallen out of fashion among American elites and out of favor among American investors?

The reality is that the United States will never again be a bigger manufacturer than China. Its much smaller population, the higher wage and standard-of-living expectations, and the dollar's status as a global reserve currency make that harder. On a practical level, it is difficult to imagine that Americans can tolerate the work habits of people in Shenzhen or Henan: working on assembly lines for eight hours a day, eating at cafeterias at designated times, crammed six to a dorm room at night. Manufacturing workers in the Midwest like to drive their pickup trucks home.

Everything starts from the recognition that something has gone quite wrong in US technology. Too many people have argued away the strategic importance of manufacturing. And the solution has to involve reconstituting its communities of engineering practice that prioritize process knowledge. It means attempting to build up every segment of manufacturing: training workers and creating incentives for manufacturers in order to relearn mass production.

This scenario sounds a bit fantastic, but if the iPhone were built

in the United States rather than Shenzhen, then an American city—
say, Detroit, Cleveland, or Pittsburgh—might be hailed as the hardware capital of the world. The follow-on innovations in consumer
drones, hoverboards, electric vehicle batteries, and virtual reality
headsets could have sprung from American firms. Engineers wouldn't
have to fly from Cupertino across the Pacific to reach their giant factories. They could iterate on product improvements closer to home,
labeling their newest products "Designed in California, Assembled
in Pennsylvania."

The United States must regain, at a minimum, the manufacturing
capacity to scale up production that emerges from its own industrial
labs. If it does not, continuing to value scientific breakthroughs rather
than mass manufacturing, then it might lose whole industries once
more—as it did by inventing the solar photovoltaic panel but relying
on China to produce them. The United States likes to celebrate the
light-bulb moment of genius innovators. But there is, I submit, more
glory in having big firms making a product rather than a science lab
claiming its invention. Otherwise, US scientists would once again
build a ladder toward technological leadership only to have Chinese
firms climb it.

Shenzhen, one day, will lose its gleam. Perhaps that process
has already started. On my last visit there in 2021, I passed by the
Huaqiangbei electronics market, where vendors were selling more
cosmetics than cables and capacitors. Hardware has become too com-
moditized a business, forcing the entrepreneurial folks at Huaqiangbei to turn their attention to China's growing demand for skin-care
products. It's hard to imagine eye creams are in line with Xi's goal to
resist deindustrialization. And yet, there it is, in a trend that a state
media headline captured as "Huaqiangbei Trades Computer Chips
for Lipsticks."

Was it an indication that not even the engineering state can resist
consumer demands, yielding to the onrush of time? The moment was

a brief one. Huaqiangbei returned once more to selling mostly electronics, as the tidal wave of Chinese industrial products is now washing up against the rest of the world. Overinvestment and an insistence against deindustrialization has protected China, for now, from suffering the unhappy fate of the American Rust Belt.

China would be better off if engineers confined themselves to building in the physical world. But they have been more ambitious than that. Beijing is made up, unfortunately and fundamentally, of social engineers. One of the major threats to China's tech power—and its global position more broadly—is the result of a disastrous decision undertaken decades ago to engage in population engineering.

CHAPTER 4

ONE CHILD

THE PURSUIT OF POPULATION control forged the essence of China's modern engineering state. Through the 1980s, Deng Xiaoping and the leadership in Beijing decided that promoting engineers into the central government was a counterstroke against Mao's misrule. They were gripped, however, by a misbegotten scientism, which used straight-line projections to predict catastrophe if China did not diminish its population. The engineering state's pursuit of the one-child policy produced more social pain than any of its other policies over the last half century. And as the state attempts to reverse its effects, it is once more employing the tools of social engineering.

In the fall of 2013, Xi Jinping gathered the leadership of the All-China Women's Federation around him at the Communist Party's headquarters in Beijing. Xi had ascended to China's highest office a year before. Looking relaxed and genial, wearing the party's standard working uniform of a zipped-up windbreaker, he told the party-linked organization that officially represents women's issues that China's economic development depends on equality between sexes. Achieving it would enable "hundreds of millions of women to shoulder greater responsibilities." The leadership of the women's federation listened intently while taking notes as they sat around him.

Ten years later, Xi addressed a new round of the federation's leadership. He'd lost a bit of weight, and his hair was grayer, but much else was the same: Xi wore the same workwear and sat in the same room, in which listeners intently took notes. Though he still wore his genial smile, his speech carried a steelier undertone. Rather than encouraging women to seek self-realization in economic development, he advised them to build families.

The vision Xi laid out to the women seated around him in 2023 sounds rather traditionalist. A woman's role is to keep the husband happy and the elders cared for; most important of all, she should have kids. "We should," Xi said, "cultivate a new culture of marriage and childbirth." That means imposing the party's doctrine on "how young people should view love and marriage, having children, and building a family." The *Economist*'s headline on the meeting was frank: "China wants women to stay home and bear children."

Earlier in 2023, China announced its first population decline since 1960 (the year millions starved from Mao's Great Leap Forward). The population drop was slight. But it was the start of a dip that will yawn larger each year for decades. By 2100, China's population is projected to halve to seven hundred million. Childbearing is collapsing in China. The country's official (and certainly overstated) number of new births has undershot even the most pessimistic projections. In 2019, China had fifteen million births; four years later, it fell to nine million. The number was below what the United Nations described as a "low-fertility scenario" only a few years before. Six million Chinese married in 2024, half the level of a decade ago. Chinese families now have a lifetime average of 1.0 children, far below the 2.1 children needed for a stable population.

In May 2023, Xi has shoved aside political convention to hang on as China's top leader for a third term. While doing so, he wrecked another norm: excluding women from the top leadership of the Communist Party. For decades, the Politburo has had at least one woman serving in the twenty-five-member group. She was often given the party's toughest tasks: Wu Yi managed negotiations for acceding to the World Trade Negotiation and handled the 2003 SARS outbreak; Sun Chunlan oversaw the enforcement of lockdowns related to Covid. Both Wu and Sun stood out for their abilities in a field of sometimes mediocre men. For his third term, Xi shrank the Politburo to twenty-four members, dropping the one space that had been given to a woman. By locking women out of China's political leadership, Xi might well have been trying to set an example.

The female body is now a fixation of the Politburo's all-male political gaze. Xi's administration has overseen a crackdown on homosexuality in China in addition to his campaign to impose traditionalism on childbearing. It's not the first time that fertility was politicized: Mao Zedong promoted births because he believed it would deter imperialist invasion. It's not the second time either: Deng Xiaoping implemented an infernal system of population control. Population engineering has now seesawed a third time, back to birth promotion under Xi.

• • •

MAO ZEDONG WAS NOT an engineer. He was a librarian at Peking University who then helped found the Communist Party, after which he became a warlord. After he established the People's Republic in 1949, Mao's stature became nearly godlike. He spent much of his time reading literature and philosophy, leaving the details of running the state to technocratic deputies like Zhou Enlai, Deng Xiaoping, and Chen Yun. Mao's gifts in military leadership as well as poetry collided in a folksy slogan he was fond of repeating: *Ren duo, li liang da*. With people come power.

In 1949, China was the world's most populous nation. After

decades of warfare, the new state didn't know how many people were within its borders. Officials guessed that China's population might be around five hundred million people. When the 1953 census counted nearly six hundred million, it was mostly a cause for celebration.

Mao viewed a big population as a source of strength. He had spent nearly half his life as a military leader fighting Nationalists and Japanese. Only a year after proclaiming the new communist state, he sent troops into Korea, mostly to fight US forces who were newly armed with nuclear weapons. Various world leaders were taken aback by his serene attitude toward atomic attack. In 1954, Mao boasted to Jawaharlal Nehru that he did not fear a nuclear strike by the United States. The imperialists, he declared, simply wouldn't have enough bombs to annihilate the hardy Chinese people. Three years later, he told a stunned Nikita Khrushchev, "We shouldn't be afraid of atomic missiles. No matter what sort of war breaks out, conventional or thermonuclear, we'll win." Mao declared he was ready to lose half of the population to fight imperialists. "The years will pass, and we'll get to work producing more babies than ever before." Khrushchev later cut off Soviet support to China's nuclear program, in part out of alarm for Mao's casualness toward apocalypse.

Karl Marx had criticized Thomas Malthus's work on overpopulation. Mao, following Marx's lead, thought it was absurd that a country could have too many people. "It is a very good thing that China has a big population," he wrote in 1949. "Even if China's population multiplies many times, it is fully capable of finding a solution. That solution is production. The absurd argument of Western bourgeois economists like Thomas Malthus that increases in food cannot keep pace with increases in population was not only thoroughly refuted in theory by Marxists long ago but has also been completely exploded by the realities in the Soviet Union and China."

Not all the other state leaders agreed. While Mao pondered literature and philosophy, Deng Xiaoping had an economy to centrally plan. Deng and other state leaders decided that five-year plans were too difficult to execute if the state could not control population. They were able to prevail on Mao to accept a few family planning policies. Through the 1970s, Mao authorized a birth control policy that included a series of incentives and fines, promoting later marriage and greater contraceptive access.

But Mao was also temperamental. Sometimes he listened to others; other times, he writhed against their restraints. Before Mao launched the Cultural Revolution, China's population surpassed seven hundred million. The continuous agitation that Mao set in motion wrought a decade of political convulsion. At the Cultural Revolution's peak, groups of workers battled over leftist doctrine, mobs pummeled people they declared to be counterrevolutionaries in mass rallies, and most schooling and work ceased so that people could heed Mao's calls to revolution. The turmoil ended after Mao's death in 1976. By then, the country was in shambles.

Among the victims were the preponderance of basic government functions. The Cultural Revolution had made a mockery of anything that could be as organized as a national census. Deng Xiaoping, Chen Yun (the most senior official on economic policymaking), and other top leaders knew that China's population was large, but they were in the dark about actual numbers. The leadership guessed that the population might have surpassed nine hundred million. When statistical authorities estimated that the population numbered nearly one billion people at the end of 1978, the leadership reacted with shock. No longer was a big population a cause of celebration. So many hungry mouths threatened to overrun Deng's modernizations.

One of China's most remarkable engineers offered a solution that sounded supremely rational. Song Jian was a missile scientist who spoke the language of mathematics and control theory. His proposed remedy was the one-child policy.

Song Jian was a man of considerable girth, his bulbous nose

framed by full jowls under a combover. At academic conferences, in which Song often gave the keynote, he spoke with a high-pitched lisp, his remarks punctuated by smiles and energetic sweeps of his meaty hands. If Song looked smug, he had reason to feel self-satisfied: Few other scientists have had their arguments embraced by China's top leaders. In political influence, Song Jian might be comparable to Albert Einstein, whose letter to the White House inspired the United States' pursuit of the atomic bomb.

Song was born in 1931 to a rural family in Shandong, a northern province that is China's second most populous. During Song's childhood, the Imperial Japanese Army landed in Shandong, which would suffer some of the worst devastation of the war. Song grew up in an occupied zone and joined the Communist Eighth Route Army as a teen, serving by day and attending school at night. Song was the only person in his high school who was able to earn a spot to university. In 1953, he earned an even rarer opportunity—to study in the Soviet Union.

At Moscow State University, Song was exposed to the thrilling new field of cybernetics. This mathematical discipline was one of several new fields, including operations research and computing, that grew from research produced during World War II. Norbert Wiener's 1948 book *Cybernetics* became a hit, not because it was filled with equations but because of its intoxicating subtitle: *Control and Communication in the Animal and Machine*. Its central idea was to develop the mathematics to control complex systems by feeding the system's outputs back into its algorithms as a continuous optimization. It is the study of regulation and control of technological or biological systems. Cybernetics has occupied an intellectual sweet spot: electrifying in its premise—attracting subordinate terms like "machine intelligence" and "systems analysis" that are irresistible in themselves—and constructed with an inherent vagueness that affords it the theoretical space to wriggle out of refutation. It is a concept that might go dormant but never completely falls out of fashion. The 1956

Dartmouth Conference coined the term *artificial intelligence* partly in reaction to cybernetics; Martin Heidegger claimed that philosophy was dying, and cybernetics would be its successor.

After Soviet and Chinese relations fell apart, Song returned to Beijing in 1960. He remained fascinated with cybernetics for the rest of his life. In Beijing, Song was appointed one of the chief scientists at the Seventh Ministry of Machine Building, the state agency in charge of rockets, where he helped to develop China's submarine-launched ballistic missiles.

Song wasn't just a gifted scientist; he also possessed a keen sense of how to maneuver for political influence. Song fell under the tutelage of Qian Xuesen, the country's best-known scientist, who was expelled from the United States and then helped to develop China's nuclear weapons. Song worked on missile guidance systems by day and on a textbook with Qian called *Engineering Cybernetics* by night. He was well known enough to have his home ransacked during the Cultural Revolution. When students accused Song of espionage (due to his occasional exchanges with foreign scientists), an alarmed Zhou Enlai packed him and other elite scientists off to China's satellite launching base in the Gobi Desert for their protection.

Military scientists like Song Jian constituted a politically privileged class under the socialist regime. Rather than being forced to make revolution, the state empowered them to build bombs and missiles. The Communist Party treated military scientists with greater deference than social scientists, whose pronouncements on economics or sociology frequently ran afoul of Mao. Over the 1950s, the chairman had bullied without mercy an economist who advocated for population control. Military scientists were also politically better connected than most university professors, who couldn't count on being heard by top party leaders. Song's privileges included engaging in scholarly exchange with parts of the outside world, as well as access to one of China's few advanced computers. He and other military scientists had political license to stomp into whichever intellectual realm pleased them.

At that moment, the world was gripped by anxieties over environmental doom. Natural scientists like Paul Ehrlich (coauthor of *The Population Bomb*, 1968) and organizations like the Club of Rome (which published *The Limits to Growth* in 1972) explained that as the global population exceeded the planet's "carrying capacity," humanity was on track to experience something between the gradual decline of living standards and the total extinction of human life. Western scientists fretted in particular about China and India, which were populous and poor. Song was still designing missiles when Mao's death cleared the way for discussion of population controls.

An overseas trip to hear from environmental doomers convinced Song that China needed radical measures to control population. In 1978, he flew from Beijing to Helsinki to take part in a cybernetics conference, where he listened to the fashionable views of natural scientists who warned of catastrophe, including presentations to determine which year the apocalypse would descend. Song later wrote that he grew "extremely excited" as he listened to these remarks.

When he returned to Beijing, he rustled up a few other scientists from the missile ministry to study population. It was a challenge because China's previous census was completed in 1964, and nobody had any real idea how large the population was. They had to rely on imprecise demographic extrapolations, eventually landing on two determinations:

- First, that if China's population growth was left unchecked (at the rate of 3.0 children per woman), then the country would have three billion people by 2050 and over four billion by 2080.
- Second, China's natural resources implied an optimal population size. Song had plugged different variables into his

ONE CHILD ★ 103

systems analysis calculations: acreages of China's arable land; the amount of water; long-term trends in the expected growth of agriculture, industry, and services. The model's results concluded that China's optimal population was no more than seven hundred million.

Today, these propositions read as bunk. Everything about them was flawed. Song wrote, "China's population by the second half of the next century would go up to 4.5 billion, equaling the total world population today. And it would continue to grow forever." Only an engineer might have believed in this sort of straight-line analysis, as if population can grow at an unvarying rate. Song had no awareness that fertility rates might fall as economic growth and educational levels rise—as neighboring East Asian countries had already realized. He presumed that China had a fixed stock of resources, leaving no room for the possibility that technological change, or Deng's pivot away from the planned economy, could increase agricultural productivity. Ironically, this mechanistic thinking made Song a bad cybernetician because his model failed to be dynamic to feedback.

In any other setting, these calculations might have been brushed aside as an unserious exercise. But the time was the 1970s, when China's top leadership didn't need foreigners to tell them that the country was facing economic stresses. The place was Beijing, where Deng Xiaoping and Chen Yun imagined that the Four Modernizations would save China, if only it followed the science. And the scientist was Song Jian, who was known and trusted by the political establishment. When Song assured China's leadership that population trajectories could be as firmly controlled as missile trajectories, they listened.

The anthropologist Susan Greenhalgh traced Song's influence on the one-child policy in her remarkable book *Just One Child*. During policy conferences, Song and his team of elite scientists made their case with calculations from China's most sophisticated computers.

Skeptics of a one-child policy were making population projections with the aid of an abacus or a handheld calculator. Song Jian presented his group's projections in precise, machine-generated lines on graph paper; other groups drew uneven squiggles by hand. It was never a fair fight. The military scientists outclassed their intellectual opponents in every possible way.

China would probably have imposed radical population controls with or without Song Jian. At the end of the 1970s, its leaders thought some form of control was necessary. Song made the scientific case that China could permit couples to have no more than one child. A few groups had murmured objections: local party secretaries who understood it would be intolerable to rural folks; social scientists who pointed out how it would create problems for retirement; and the army, which worried about recruitment.

They lost. On Song's side was the formidable Chen Yun, who pushed hard for a one-child policy. The rest of the leadership mostly agreed as well. Only a few policymakers wondered whether it might not be better to permit two children per couple and whether education and greater contraceptive access might not be sufficient. Deng Xiaoping's decisive voice weighed in favor of only one child. He and Chen Yun were seasoned administrators who had an intuitive understanding that targeting *one* child provided simplicity to the millions of local officials responsible for enforcing the rule. Song Jian pushed on an open door. His projections allowed them to believe that the crudest goal was also the most necessary.

Beijing adopted the one-child policy in 1980.

Song's authority fit with Deng's goals to cast his new policies as a modernizing, scientific force, with the precisely drawn graphs to prove it. Deng and Chen Yun were starting to think about China's growth in per-capita terms, which pushed them into the faulty line of thinking that resources per person were higher when there were fewer people. Years later, Song gloated about how much smarter natural scientists were than social scientists. His strategy, if he were ever attacked, was to "withdraw into the sanctuary with the high prestige of natural science." Song never stopped congratulating himself. In a 1988 book, *Population Systems Control*, he and a coauthor wrote, "Using statistical and quantitative research methodologies, population studies have been freed from the interference of human emotions and the damaging effect of popular ethics."

The Communist Party, however, could not fully ignore human emotions and popular ethics. It knew that the people would react to this policy with incredulity. In an approach almost without precedent, the Communist Party published an open letter addressed to all members, asking them to set the example of having only one child. According to Greenhalgh, the propaganda authorities gave Song Jian the honor of writing the first draft. The result went as well as could be expected. Song was too arrogant to address the people with tactfulness, so officials threw out his draft and gave the task to propaganda professionals.

The open letter of 1,600 words was published in the *People's Daily* that September. "In order to keep China's population below 1.2 billion by the end of this century," it began, "the State Council has issued a call to the people of the whole country, advocating that each couple should have only one child. This is a major measure related to the Four Modernizations, to the health and happiness of future generations, and to the long-term and present interests of all the people. The Central Committee requires all Communist Party members to take the lead . . . to actively, responsibly, patiently, and meticulously carry out publicity and education to the masses."

This letter adopted a plaintive tone. It "advocated" for couples to have only one child. It took pains to sound reasonable, citing the stagnation of living standards and the stress that the population was putting on farmland. Even today, the name of the policy barely evokes the violence involved with its implementation, in which posses of enforcers reached their hands into a woman's most intimate parts in order to carry out, at times, forced sterilizations and abortions. Enactment of the one-child policy meant forcing a mostly rural people to change deeply ingrained habits. It was social and population engineering at scale.

. . .

THE ONE-CHILD POLICY BEGAN as a shock campaign and matured into a labyrinthine administrative apparatus. Over the thirty-five years of the policy's existence, it left few Chinese families untouched. By 1990, in order to have a first child, a woman needed up to twelve documents from her workplace and various party officials and a consent form agreeing to contraceptive measures after birth. The less fortunate were caught in the mass sterilization and abortion campaigns that swept through the countryside. For rural families describing what it was like to live through those times, "wrenching" becomes the descriptor of first resort.

Beijing designated Qian Xinzhong, a former general of the People's Liberation Army, to be the head of the State Family Planning Commission. Qian planned the opening phase of the enforcement as carefully as a military campaign. He called on roving teams of family planning officers to be "shock brigades" who must implement "man-on-man tactics" in the great battle for family planning. Crucial in his conception was the "shock attack," a term from socialist campaigns emphasizing political mobilization to achieve decisive results. These teams consisted of state and party cadres, local enforcers, and a medical team that would traverse villages. Hospitals had to be prepared to carry out the "four procedures": IUD insertions, tubal ligations, vasectomies, and abortions.

Qian threw these shock troops against the bewildered masses of China's rural folk. When the one-child policy started in 1980, urban fertility rates were already trending toward 1.0 child per couple, while rural fertility was closer to 2.5. For the four-fifths of Chinese who lived in the countryside, having several children was the basis of economic security. Without multiple children, and ideally sons, a farmer couldn't count on having enough work and old-age support.

In 1982, China was finally organized enough to undertake its first census since 1964. Deng and Chen regarded the results with glumness. China's population increased by three hundred million in those eighteen years, becoming the first country ever to surpass one billion people. The leadership felt even more convinced of the need for population control. The Communist Party had declared family planning a "foundational national policy" and wrote it into the constitution, removing it from the realm of debate and empowering Qian's most ruthless instincts.

In 1983, Qian mobilized party and state offices at every level for a big push. That year, the state sterilized sixteen million women and carried out fourteen million abortions. By comparison, in the pre-policy year of 1975, the state performed only three million sterilizations and five million abortions.

Hitting these numbers required escalating coercive tactics. The first measure in the official toolbox was browbeating. Local officials would visit pregnant women as part of "persuasion groups." This posse of up to ten men seldom appeared as sweet-tongued advocates. One American academic witnessed a group of women in Guangdong separated from their husbands and sent to the village hall. There, they were given unceasing lectures to give up their pregnancy for the good of the country, and then were called upon one by one to give their consent to an abortion while being prohibited from returning home until they had done so. A 1982 *New York Times* report quoted a family planning official from Guangdong saying, "On average, each person takes 10 times to be persuaded. The most difficult person can take up to 100 times." The piece also cites women hauled before mass rallies and harangued into consenting to an abortion.

Slogans exhorted cadres not to slacken their work. "Any method that reduces fertility is a good method," said one. "Take all measures and overcome difficulties with creativity," said another. These were tantamount to offering open license to take any means necessary to terminate a woman's pregnancy. The browbeating often worked: Few families could endure up to a hundred visits by a rotating cast of officials with ever more insistent demands. But if the tactics still were ineffective, the officials could threaten firing or fines worth up to several years of wages. They could detain the woman or a family member, which required paying for one's food each day without being able to contact the outside world. Sometimes they carted off furniture or sewing machines, seized cattle and other livestock, or sent a bull-dozer to tear the roof off their home. A family that thought it had the means to support an additional child then had to ponder whether it still could do so.

Nothing was more important than hitting numerical quotas. Local officials received cash bonuses and good reviews if they met their sterilization and abortion quotas; if they did not, they saw their pay docked and were demoted. Enforcing family planning was part of an official's personnel evaluation. Nicholas Kristof, then a reporter based in China, wrote about a woman who was seven months pregnant when officials demanded that she give birth right away. These officials formed a shock brigade to round up all third-trimester women because they had some birth quotas left in the year, while they weren't sure whether they would have many next year. Against the objections of the woman's doctors, they induced an early birth. Kristof described how she nearly hemorrhaged to death during the birth. Her child died. And this mother-to-be was left physically disabled.

If a woman was still not persuaded, then officials might carry out a forced abortion. Often, they operated in the third trimester because the woman could no longer conceal her big belly. In some cases, a baby came out alive. Michael Weisskopf, who reported on the one-child policy in a series of pieces for the *Washington Post* in 1985, wrote that doctors sometimes injected formaldehyde into a baby's head or crushed the skull with forceps. More typically, doctors would smother the newborn or leave it to die of exposure.

Song Jian's home province of Shandong experienced the most notorious incident of strict enforcement. Zeng Zhaoqi, newly appointed party secretary of Guan County, was humiliated that it ranked last in the province for family planning. So he summoned the twenty-two most senior party officials one day in April, berating them for their failings and shouting that their measures must be more extraordinary. He demanded there be zero births in the county between May 1 to August 10. In reports now censored, residents said that every woman was forced to have an abortion, no matter how far she was into her pregnancy or whether it had been authorized. Zeng found toughs from other counties—since locals were reluctant to hurt their own—to halt births.

This incident in Guan County is known by two names: the "childless hundred days" as well as the "slaughter of the lambs," since 1991 was the year of the sheep in the Chinese zodiac. The slaughter ended well for Zeng. He was rewarded with successively more desirable promotions in Shandong. His superiors didn't seem to have a sense of irony when they appointed him later to be the deputy head of the provincial committee on Caring for Future Generations.

Though Qian Xinzhong didn't hesitate to order late-term abortions, his preferred tool was sterilization. Abortions were messy and traumatic for all; sterilization was simpler to carry out and could represent a decisive solution. Doctors might automatically implant an IUD immediately after a birth, sometimes without bothering to inform the patient. Since women attempted to remove these implanted rings, Qian preferred tubal ligation—an irreversible procedure. He advocated for universal sterilization of couples who already had two children. That wasn't implemented everywhere, although there are reports of maternity wards sterilizing mothers immediately after a second birth. Automatic sterilization was a step that provoked unease in Beijing. Since infant mortality was still high, rural families feared the permanent loss of the ability to have a child. But by 1999, China's health ministry statistics show that 35 percent of married women of reproductive age had been sterilized.

The campaign produced agony for rural folks. They fumed that the state was treating them exactly as they dealt with their own livestock. Wives and daughters were being sterilized in much the same way that farmers spayed their pigs. It didn't help that the abortion posses sometimes literally carted women off in hog cages. Weisskopf wrote in the *Washington Post*, "Expectant mothers, including many in their last trimester, were trussed, handcuffed, herded into hog cages and delivered by the truckload to the operating tables of rural clinics." The toll on women's bodies was enormous. The stainless-steel IUD rings inserted after births created long-term physical problems, provoked menstrual bleeding, and tended to wear out after two years. Abortions and invasive tubal ligations were often done in a hurry and en masse, sometimes without anesthetic. Men could have volunteered for vasectomies. But typically, four women received a tubal ligation for every vasectomy.

The one-child policy didn't create so much difficulty in urban areas. Many people in cities were able to navigate the situation. They were also more likely to have the means to travel abroad to have a second birth. And the party trod cautiously in restive minority regions, since problems associated with overpopulation were caused by the majority Han settlers in Tibet or Xinjiang, not by the locals. Sometimes, villagers were able to pay the fine for an additional child and get on with their lives. A bribe could do the trick. Local officials had an interest, after all, in concealment, protecting both themselves and their villagers.

When people needed to resist the onslaught, they wielded what

James C. Scott called weapons of the weak. The most straightforward means of resistance was to escape to a different village. A mother might return with a newborn and hope for leniency with a fait accompli. But it was a risky strategy to produce an out-of-plan child. Many jurisdictions did not allow them to have the schooling or medical benefits available to an authorized birth. It meant they might miss early inoculations, be barred from school enrollment, and experience forfeiture of their land rights. They were essentially second- or third-class citizens whose most likely fate was to become unskilled migrants.

Women tried to time their pregnancies so that they would give birth in winter, when they could bury their growing belly under layers of heavier clothing. Officials knew they were not detecting all the pregnancies, so they offered financial rewards for neighbors to snitch. Since China's minority groups enjoyed some leniency to have more than one child, people discovered Tibetan, Dai, Miao, or some other such ancestry that they had previously forgotten to disclose to authorities. After Beijing loosened the policy permitting families to have a second child if their first was disabled, the writer Peter Hessler discussed the story of a family that rented a disabled child they claimed was theirs in their application for a second birth permit (which was successful).

Confronting birth-planning officials was a tactic of last resort. Rural folks said these officials were after only three things: your money (through fines), your grain, or your life. Enraged villagers sometimes retaliated against officials by destroying their homes or livestock. Kidnapping the children of shock brigade leaders became a common fantasy, sometimes executed. Arson was such a common revenge that cadres developed a phobia of fire: Ten days after one person was promoted a "tubal ligation team leader," his home was razed to the ground. Attacks against birth-planning officials became so frequent that some areas drew up laws specifically to prohibit retaliation: Shaanxi, for example, passed a law against "insulting, injuring, or slandering birth planning personnel or their families." The government eventually came up with a special insurance scheme for covering accidents and damage to the homes of birth-planning officers.

They became some of the most hated people in Chinese officialdom, but these enforcers had little autonomy and few privileges. Only half had completed high school, and only one in eight received any medical training, even though many of them were thrown into doing invasive procedures. Most were poorly paid and developed poor morale from being treated with contempt by other officials while they implemented the one-child policy. In three separate state surveys, more than half of birth-planning officials expressed a desire to quit their work.

The zenith of the one-child policy was Qian's big push of 1983. Later that year, he lost his job. Beijing then loosened the policy slightly, releasing new guidelines to repudiate shock tactics and permitting more couples to have a second child, especially people in rural areas. The brutality, however, continued. China's health yearbooks reveal another high tide of sterilizations and abortions in 1991, the year of the mass slaughter in Shandong's Guan County, as a large cohort of women entered childbearing age. After that, however, the one-child policy became less confrontational, although sterilizations and abortions remained at high levels.

For more than a decade, the one-child policy produced a campaign of rural terror. Local officials had to convince people they were serious about changing birth habits. Documentation from that time, occasionally surfacing even in state media, reports forced sterilizations and abortions, as well as public incidents of drowning newborns to make people realize that the state and its one-child policy meant business. The state was trying to enact compliance by changing cultural attitudes.

One of the notorious legacies of the one-child policy was the high rate of female infanticide. Rural families tended to have two preferences: to have multiple children, and that at least one should be a boy. The one-child policy collapsed these desires into a preference for sons. Reports of female infanticide poured into government offices. Baby girls were being smothered, drowned, poisoned, or left in trash heaps as soon as they were born. "At present, the phenomena of butchering, drowning and leaving to die female infants and maltreating women who have given birth to female infants have been very serious," state media forlornly admitted. "It has become a grave social problem." By the early 1990s, ultrasound machines were in widespread use, permitting parents to engage in sex-selective abortions. That meant fewer killings after birth. But it didn't stop China's official sex ratio at birth to reach 120 boys born for every 100 girls in 1999. That ratio has since declined to 111 boys to 100 girls. In the intervening decades, however, demographers estimate that around forty million women are "missing."

Not every family had the heart to give up their newborn daughters. Kay Ann Johnson was a professor from Massachusetts doing fieldwork in northern China when she adopted a three-month-old girl. Two decades later, she wrote a moving book, *China's Hidden Children*, in part, as she put it, to help Chinese children who were adopted abroad understand the impossible circumstances of their birth parents. It dawns on some out-of-plan or adopted children to say, as early as age three, "I should never have been born" when they learn of their own legal discrimination or abandonment. In scores of interviews, Johnson found that rural families, including the father, experienced lasting emotional scars from anguish and rage that they could not keep their child. They felt that they had no choice in the matter. But they also suffered a deep sense of loss and personal failure.

When birth parents abandoned (almost invariably) their baby girl, they tried to find a good adoptive family, usually a childless couple or a family already with multiple boys. After depositing the girl at the doorstep, they might set off some firecrackers for attention. Anyone emerging out of their doorway would glance down at a newborn and immediately understand the task asked of them. If birth parents couldn't identify a good family, they might abandon the girl in the city, reluctantly, since they had little idea who might pick up their child. It became a common story for city folks to hear a baby's cries from inside a cardboard box or by a trash heap.

The one-child policy increased child abandonments and child abductions. Trafficking rings stepped up to mediate between families who could not keep their child and families who wanted another. Sometimes, they abducted girls to fulfill demand for future brides; most of the time, they abducted boys because more families wanted sons. Child smuggling became an interprovincial venture. In 2004, twenty-four baby girls in tote bags were found on a long-distance bus, drugged to keep them quiet, bound for adoptive families. That led to the bust of a large baby-trafficking ring whose leaders were sentenced to death. Police raids to rescue trafficked children continued for much of the 2000s.

Johnson recounted several instances of forcible seizure of children by the state. In one account, seven men descended from several directions on the home of a family with an out-of-plan child: "The government had taken their baby, stripped them of their parental rights, and left them heartbroken and powerless to do anything about it," Johnson wrote. "It had been nothing short of a kidnapping by the government, leaving them no recourse."

State-enacted kidnapping was one of the perverse consequences of the one-child policy. China started sending children abroad starting in the early 1990s. Adoption agencies sprang up for American families who went to China to bring home a child. Though the process of vetting foreign parents was rigorous, the procedures for making babies available to them was not always transparent. Orphanages didn't always treat children well: One American on an adoption trip to

Wuhan wrote that a family on his trip received two successive notices that their designated adoptee had died. International conventions agreed to by Beijing required adoptive parents to offer a donation. The size of the required donation was between $3,000 and $5,500, which was an enormous sum for any Chinese orphanage. That created a perception that orphanages were in the business of selling children abroad. This label was often not fair. Unfortunately, local governments sometimes really sought to benefit from these big payments.

Hunan, Mao's birthplace and the province where many of the worst abuses of the one-child policy have been reported, distinguished itself on excess. Parents in rural Longhui County reportedly grabbed their babies to find hiding places whenever family planning officers showed up. Officers snatched at least sixteen children who didn't have proper papers and placed them in orphanages. Eventually, a few ended up in the United States, Poland, and Holland. Longhui residents have accused the government of abducting their children for revenue, raising the horrifying possibility that American adoptive families may have taken in children who weren't actually abandoned.

The one-child policy persisted into the era of online virality. In 2012, Feng Jianmei, a twenty-three-year-old mother in rural Shaanxi, was pregnant with a second child. When she failed to pay a fine demanded by birth-planning officials, they shoved her into a van, blindfolded her, made her sign a document she could not see, and gave her shots to induce a stillbirth. That in itself might not have been remarkable. What was unusual was that Feng's husband uploaded a photo of her—exhausted and with her bloody, stillborn fetus lying beside her—on China's nascent social media platforms. When the post blew up, younger people reacted with outrage. One commenter stated that the family planning system has been "openly killing people for years in the name of national policy."

Beijing took too long to end the one-child policy in large part due to bureaucratic inertia. The State Family Planning Commission had over 500,000 workers, 1.2 million local enforcers, and 6 million village officials engaged in enforcement. It collected $200 billion in fines over its lifetime, according to state media. For the millions of people given jobs by this bureaucracy, it was worthwhile to keep the policy from expiring. And the commission kept finding evidence that families were hiding their out-of-plan children. It wasn't until the 2010 census conclusively proved the fertility rate had collapsed that the central government dissolved the commission.

China ended the one-child policy in a desultory process, formally terminating it after the bureaucracy stopped putting up a fight. The one-child policy became a two-child policy in 2015, then a three-child policy in 2021. Over the thirty-five years of the one-child era, China performed a total of 321 million abortions (not far off from the present population of the United States) and sterilized 108 million women and 26 million men. In 2024, Beijing announced that it would end international adoptions. By that time, more than 150,000 children had been sent abroad (around half to the United States), almost all girls.

• • •

AS I WAS WRITING this chapter in 2024, my wife, Silvia, suffered a miscarriage. It was in the first trimester of our first pregnancy. As we grieved, I returned to writing about these mass sterilization and abortion campaigns. If anything, it became more difficult to imagine how the state dragged away so many women to forced abortions in their third trimester. Meanwhile, women in the United States were fretting about curtailments over their reproductive rights. Neither forced nor prohibited abortions are humane, Silvia and I felt, which means the state should leave families, and especially women, with a choice.

I was born in 1992. When I spoke to my mom about the one-child era, she remembers the bureaucracy more than anything else. She needed to fill out a lot of forms to have me, including committing to contraceptive measures after my birth. She was surprised when I told her that China recorded the second-highest number of abortions the year before my birth (fourteen million, a few hundred thousand shy of the peak enforcement year of 1983). Since my parents were urban residents, they didn't feel the brunt of this enforcement, which fell on the countryside. They also had the fertility preferences of urban folk, which tended toward one child. My parents discussed having a second child after we moved to Canada when I was age seven. But they didn't feel strongly about it, so they didn't.

The one-child policy left subtle imprints among urban folks. Chinese people my age rarely ask each other whether we have siblings; it becomes quite curious if someone does have any. I have three cousins, and my family encourages me to refer to them as sisters in order to create closeness.

Time has worn away some of the memories of the traumas. But they are still there for rural folks. Foreigners curious about the one-child policy will probably not hear vivid stories from the Chinese they speak to, who tend to be relatively privileged people from cities. It is rare for rural folks to be able to study and live abroad. They sometimes have a hard time even moving to cities, given the restrictions of the *hukou* system—another social engineering project—meant to restrict internal migration. The one-child policy is another reminder of a phrase that resonates a great deal for me: "Chinese peasants, your name is misery." It was coined by Sun Dawu, a rural entrepreneur now jailed for his advocacy.

The one-child policy could only have been formulated by the engineering state. No other country would have let a missile scientist anywhere near the design of demographic policy. Its roots lie partly in the control tendencies of Deng Xiaoping and Chen Yun, who wanted to engineer the population so that they could engineer the economy. Partly in reaction to Mao, partly using language given to them by Song Jian, they viewed themselves to be acting on a science that was detached from popular passions, based on Western ecological concerns, and formulated in terms of control theory. They understood themselves to be acting as technocrats.

The lawyerly society debated the one-child policy and rejected it. The United States and other Western countries also considered implementing strict population controls in reaction to *The Population Bomb*. Social scientists, especially economists, were quick to criticize the flaws in these linear projections. But in China, social scientists had become meek from Mao's bullying. At this critical moment, the country lacked the intellectual antibodies to resist the policy's adoption. Chinese leaders were just enough exposed to the West to absorb this neo-Malthusian doomerism, without being exposed enough to the Western pushback against it.

And the one-child policy could only have been *implemented* in the engineering state. While the state possessed a bureaucracy to enforce controls of such extraordinary scale, there wasn't a sufficiently developed civil society to fight for legal protection against it. The Communist Party is built to implement campaigns of this sort. That is what Leninist parties, which are hierarchical and mobilization oriented, do. When it put someone as savage as the general Qian Xinzhong in charge, it was able to achieve astonishing numbers of sterilizations and abortions.

The one-child policy is one of the searing indictments of the engineering state. It represents what can go wrong when a country views members of its population as aggregates that can be manipulated rather than individuals who have desires, goals, or rights.

Susan Greenhalgh related the story of Liang Zhongtang, who was one of the few vocal opponents of the one-child policy inside the party. He was, however, only a professor in the backwater of Shanxi province, making him far removed from actual policymaking. Liang attempted to make China's leadership see villagers as people, whose childbearing desires were embedded in a network of cultural values and economic needs. He handily lost the battle to the cybernetics faction led by Song, who viewed rural folks as a variable to be controlled as the state saw fit. "The size of a family is too important to be left to the personal decision of a couple," Qian Xinzhong said. "Births are a matter of state planning, just like other economic and social activities, because they are a matter of strategic concern."

For my parents, it was apparent that China was facing shortages when they grew up. They needed ration tickets for everything: rice, eggs, cooking oil, bicycles, an apartment. Obtaining almost anything was difficult. My mom and dad were among a handful of students able to earn a spot at university. When I asked my dad whether the one-child policy made sense to him, he replied, *ren tai duo*. Too many people! It's a common refrain. Anyone taking the subway during rush hour or touring scenic spots over national holidays might hear it muttered still today.

There's no question that Chinese people experienced severe shortages of everything prior to the adoption of the one-child policy in 1980. But these shortages were the result of the socialist planned economy. This system was characterized by agricultural collectivization, an emphasis on heavy industry, and lavish spending on national defense, leaving little left for consumer production. Consumer shortages eased when Deng moved China away from socialism. It's unclear if Deng was aware of the irony that he was attempting to impose planning on the population while he was trying to dismantle planning for the economy.

While China's population has increased by 40 percent since the start of the one-child policy—with Beijing doubling in size and Shanghai quadrupling—Chinese are living better than they ever have. They are rich in material possessions and can more easily access the finer things in life. That shift was chiefly produced by ceding economic freedom and allowing people to trade with the rest of the world. And though Mao's economic policies caused famine and destitution, it's hard not to agree with his remark from 1949: "Even if China's population multiplies many times, it is fully capable of finding a solution. That solution is production."

Rather than acknowledge that it could not deliver the goods, the Communist Party decided instead to blame the people. It was their "overpopulation" that was the problem, not the inadequate economic system that the leadership insisted on.

After people accommodated the one-child policy by resorting to female infanticide, Beijing felt some embarrassment over the ensuing headlines. Rather than acknowledge the impossible choices it had forced people into, the Communist Party once again blamed the people. Cadres declared female infanticide a symptom of "feudal practices" and a "peasant mentality." Any efforts to actually address the problem were half-hearted, consisting of exhortation and an educational campaign. Population control was still China's primary problem. Millions of missing girls were a distantly secondary concern.

The Communist Party invoked the environment to justify the one-child policy. Shortly after implementing it, China began its great industrialization, which lifted economic growth while ruining much of the country's ecology: polluting its lakes, pushing heavy metals into its soils, and delivering coal smoke into its air. It wasn't overpopulation that destroyed Shenzhen's oyster ecosystems; it was state-directed industrialization. The one-child policy occurred in parallel with China's wanton devastation of its environment. Perhaps the policy even offered policymakers moral license to justify environmental devastation.

How will the one-child policy be remembered? At its conclusion in 2015, a trio of demographers offered an assessment in the journal *Studies in Family Planning*: "Future generations will likely look back at China's one-child policy with bewilderment and disbelief. To many it will be incomprehensible why, of all countries that faced the challenge of rapid population growth in the second half of the twentieth century, only China went to such an extreme; incomprehensible why in a society based on respect for the family, kin, and filial piety, the government enforced a policy that effectively terminated many kin ties for at least a generation; incomprehensible why China instituted such a policy after the country had already experienced substantial fertility decline; and incomprehensible why China waited so long to end such a harmful policy."

Of all the critiques of the one-child policy, perhaps the most poignant is that it was not necessary to reduce China's fertility rate. That was already falling due to earlier, less coercive family planning policies. China's fertility rate was around 6.0 per woman at the start of 1970; a decade later, when the state implemented the one-child policy, the fertility rate had already fallen to 2.7. Professional demographers still debate the extent of the fertility decline that the one-child policy produced. Official state media have claimed that family planning measures over four decades prevented four hundred million births. That figure is marred, however, by the same sort of linear assumptions embedded in the projections by Song Jian. Any effort to determine the number of births prevented by the one-child policy is made difficult by patchy data released by the government.

Demographers give credit to Deng Xiaoping for driving down fertility not through the one-child policy but through economic reopening. Higher rates of urbanization, educational attainment, and, most of all, economic growth have been the best contraceptive measures devised by modernity. These were factors that drove neighboring Japan, South Korea, and Taiwan to lower their fertility rate too.

The true legacies of the one-child policy are psychological scars, sometimes physical scars for the mothers, gender inequity, and a rapidly aging population. An aging population was always a predictable concern with the one-child policy. In fact, it was acknowledged by the Communist Party's open letter in 1980. The letter brushed aside concerns that a future child would have to support four grandparents, saying instead that the policy would secure such national prosperity that the state would be able to afford generous pensions for all. Propaganda authorities asked people to trust the government, before switching to tell them to stop burdening the government. Whereas one of the former propaganda slogans read, "Have one child, it will be enough; the state will care for you when you're old and tough," a new slogan now reads, "Have three children so you won't have to seek state-supported elder care."

Neither Song Jian nor Qian Xinzhong appeared to have much regret about their roles in the one-child policy. Qian was given a curious honor in 1983, when the United Nations Fund for Population awarded him (along with Indira Gandhi, who presided over a campaign of forced sterilizations in India) its inaugural Population Award. He never held another office after running the family planning commission, dying in 2009 at the age of ninety-eight.

Song Jian is still around. After 1980, he held a dazzling array of high positions: president of the Chinese Academy of Engineering, minister for science and technology, state councilor, and membership in the Communist Party's Central Committee for twenty years. He was a natural politician. When China's population did not explode, he was able to declare victory anyway for defusing China's population bomb. Song never lost his enthusiasm for cybernetics. In an ambitious article written in 1984, he advocated for a strong leader, supported by teams of technical cadres, to employ cybernetics to manage the entire society.

Before Song retired in 2002, his final project involved chronology. After visiting Egypt, he grew embarrassed that China apparently lacked a detailed chronology of its ancient civilization. Though China claims 5,000 years of continuous history, the first few thousand are a bit hazy. Song established that Chinese civilization was 1,400 years more ancient than previously recorded. Another great feat! There's nothing that many Chinese love to hear more than the idea that the nation's past glories were even more glorious than anyone had grasped.

It was the last example of how Song applied his brilliant mind along with an amateur's enthusiasm to serve state ends. His work in historical chronology might be harmless, but his involvement in fashioning the one-child policy produced so much trauma. Perhaps I was wrong to compare Song Jian to Albert Einstein in terms of their influence after pressing scientific analysis into the hands of senior leaders. The more apt comparison for Song might be to Trofim Lysenko, the agronomist who aligned himself with Soviet orthodoxy and helped perpetuate famines in the Soviet Union.

Song's example is one reason that I've become suspicious of anyone who advocates "following the science." We have to get quite worried if anyone in power starts saying that science alone is an object to be pursued rather than having to situate it in a social and ethical context. There is still truth, I think, to Winston Churchill's quip that scientists should be "on tap, not on top."

• • •

BY THE YEAR 2100, China's population is on track to decrease to seven hundred million people. As it turns out, that was the optimal population size calculated by Song Jian.

Far from celebrating this decline, Xi Jinping and the rest of China's leadership are trying to reverse it. Each year after 2022 will see slightly fewer people powering the Communist Party's great odyssey toward national rejuvenation. Maternity wards are starting to shut down in several provinces since there are fewer newborns. In 2025, adult diapers are expected to outsell baby diapers. China has already grown old before it grew rich: When Japan's population started to decline (fourteen years before China's), it was more than twice as rich.

Was there ever a country that exerted so much effort to deplete its own population? Mao would be astonished by the one-child policy, as would almost any other world leader before him. With people comes power. Political leaders have universally tended to want more of both. Demographic decline will entail a slow grinding down of China's actual capabilities to achieve geopolitical preeminence.

China's low birth rate worries Xi Jinping and the rest of the Communist Party. In the 2023 meeting with the women's federation, Xi vowed that over his third term his administration "will improve and implement pro-fertility policies." The shift to the two-child policy in 2016 and the three-child policy in 2021 did not produce many more births. China's fertility rate of 1.0 is now lower than Japan's and keeps falling short of even recent low-fertility projections.

So the party has grown more vocal in blaming one final group: women.

It's hard to be a woman in China today. Many of them did not survive the one-child policy: There are approximately forty million more Chinese men than women. Though the country has plenty of successful female entrepreneurs and billionaires, Xi has shoved women out of the top echelon of the Chinese government. His primary message is that women must become docile promoters of family harmony, which means bearing more children. That theme is also being echoed by the rest of society. Rather than being joyful, Lunar New Year is an irksome time for younger Chinese women. They must face dozens of relatives, from whom they expect only one question: "When will you marry?" to the single woman, and "When will you have kids?" to the married.

The journalist and sociologist Leta Hong Fincher has documented the brazen insults that women have to endure, especially from state media. In her book *Leftover Women*, she chronicles how women tend to be discarded (often in contempt) once they've reached unmarriageable age, which state media considers to be twenty-seven years old. She documented how women must endure all manner of insulting headlines lamenting their case: "Eight Simple Moves to Escape the Leftover Women Trap," and the column posted shortly after Women's Day, "Do Leftover Women Really Deserve Our Sympathy?" The aim of stigmatizing singlehood, Hong Fincher writes, is to stop urban women from delaying marriage and childbirth much further.

Even if a woman is married, state media is unkind. A Xinhua news editorial urged women not to make a fuss if they discover marital infidelity: "When you find out that he is having an affair, you may be in a towering rage. But you must know that if you make a fuss, you are denying the man 'face.' Try changing your hairstyle or your fashion." The Women's Federation is often the amplifier of these messages. Since it is the state's designated organization on women's issues, it is often in the position of enforcing state policies. One former employee of the Women's Federation told the *Wall Street Journal* that her office in Guangzhou spends more of its budget to give to social media companies to censor gender-related topics than on women's advocacy.

As China shifts away from birth control under Deng (and several successors) to birth promotion under Xi, it is relying once again on the tools of the engineering state. But the state is starting to see that this dial cannot be turned back. Although the state has had many tools to prevent births, it can't seem to find the right tools to encourage copulation.

State media has become increasingly desperate to urge births. In 2018, two academics proposed the establishment of a "birth fund," to which all workers under the age of forty must contribute, while couples who have more than one child could apply for subsidies from the fund. Decried as a tax on the childless, that proposal went nowhere.

In 2021, an unsigned commentary appeared in a state media paper demanding that all members of the Communist Party have three children, in unusually vehement terms. "That would not only be good for the family," the editorial said, "but also national development needs. It must be every party member's responsibility to have three children! They can't offer wimpy reasons not to marry, and to have just one or two kids." This editorial was deleted after an online outcry. Demanding a politically loyal cadre to have many children is not new.

I think Heinrich Himmler, however, said it better when he exhorted SS officers to have more than four children: "Think of Bach! He was the thirteenth child in his family! After the fifth or sixth, or even the twelfth child, if Mama Bach had said 'that's enough now,' which would have been understandable, the works of Bach would never have been written."

Three decades of persuasion in favor of one child has worked too well. All women of childbearing age grew up in a China insisting that the best number of kids was one or zero. In response to social and government pressures to have more kids, women retort on social media with pictures of slogans that used to be plastered all over the countryside urging families to reduce fertility. Half of all Chinese women born after 1995 told the Chinese general social survey of 2021 that they desire one or zero children. The bullying they have to endure from the Women's Federation and state media hasn't made them enthusiastic about childrearing. When a southeastern city offered incentives for leftover women to marry rural, unemployed men, women reacted with incredulity. Why should a woman leave a city job to marry a man she regards as a deadbeat? Marriage has become even less appealing since Chinese judges are increasingly reluctant to grant a divorce: 70 percent of divorce applications were granted in the mid-2000s, a rate that fell to 40 percent a decade later.

The one-child policy persisted for one and a half generations. Its effects will echo far longer. I am skeptical that the engineering state will be successful in producing a surge of births. There have been pronatalist policies in other countries (Hungary, Israel, and many others), with little evidence that they could structurally push up birth rates for long. China is catching up with other countries in these fertility policies, held back both by technology as well as social attitudes. The country has only six hundred hospitals officially authorized to offer in vitro fertilization services. And the state makes it illegal for unmarried women to freeze their eggs. To preserve their fertility, single women have been forced to travel to Taiwan or Thailand in order to find egg-freezing services.

It is possible that China will be able to implement profertility policies, as Xi has promised, more successfully than anyone else through tactics of the engineering state. So far, however, women of childbearing age haven't been interested. Perhaps the state will invent a technological solution to produce more Chinese children. At the moment, the efforts are low tech. Women in urban cities are reporting that they are regularly getting calls from neighborhood officials asking when they plan to have children. These officials are inquisitive, asking when a woman has had her last period, and argumentative, insisting that owning a cat can be no substitute for a child. Most of all, they are nagging. One woman posted, "Government officials have asked me five or six times when I plan to have a child, while my parents have asked me only once." She goes on to say, "These officials call only to rush me, not to offer any support."

Rather than being totally fixated on women, the engineering state is now also thinking about men. State media has started to fret about leftover men too. The tens of millions of Chinese men who will never be able to find wives may become a threat to public safety, who could, in the words of one university researcher, "be driven to kidnap women or become addicted to pornography." Men have also taken to social media to complain that it's getting too difficult to obtain a vasectomy. Some hospitals turn men away from vasectomy unless they can prove they already have children. National health yearbooks reveal a breathtaking collapse in vasectomies performed in China. They fell from 181,000 in 2014 (the start of Xi's rule) to fewer than 5,000 in 2019. In the new era, men are getting a taste of birth planning too.

The one-child policy is a rebuke to the idea that the population can be so easily engineered. Social engineering in this case has pro-

duced a spiritual defeatism manifesting in broad exhaustion throughout society. Exactly four decades after China began the one-child
policy, it would enact an even more ambitious social program: from
controlling people's bodies to engineering their souls, this time with
the aid of digital surveillance.

CHAPTER 5

ZERO-COVID

CHINA'S RESPONSE TO THE Covid-19 pandemic embodies all of the engineering state's merits and madnesses. It is a powerful reminder of how the engineering state could accomplish things that few other countries would even attempt, while revealing how its literal-minded enforcement can lead to tragic results for human well-being and freedom. I lived through all three years of the zero-Covid strategy that China pursued to stomp out the highly transmissible virus. In the first year, the country felt like a realm of serene calm after it pushed out the virus that was raging far away. In the second year, it still felt pretty good, though all of us were getting antsy as we wondered how the government would organize its exit from the policy. In the third year, everything went to smash.

In 2020, at the end of the first year of the Covid-19 pandemic, I moved from Beijing to Shanghai. I was driven away as much by the intensity of Beijing's political temperament as I was drawn by the splendor of Shanghai's commercial character.

Beijing had been China's seat of empire for centuries when

Stalinist architects began, in 1949, to reshape the city for socialist magnificence. Visitors from Shanghai liked to tease those of us
in Beijing: "Why would you live in Pyongyang when you can live in
Paris?" It was annoying. Then the SARS-CoV2 virus burst out from
Wuhan. Pandemic regulations made Beijing an even more tightly controlled city than in normal times. When I heeded these exhortations
and made the move to Shanghai, which imposed substantially looser
restrictions, life really did feel cheerier. People were walking the
streets, many of them unmasked in Shanghai's considerably warmer
clime, out and about having a great time.

Before Soviet-trained engineers refashioned Beijing for monumentalism, the French built Shanghai for pleasure. Colonial powers
transformed Shanghai from a modest trading port in the nineteenth
century into the beachhead for foreign powers to penetrate the country's giant market. The British, the Americans, and the French each
carved out enclaves where their residents could disregard Chinese law.
The second-largest bank house in the world was built in the British
and American zone, alongside insurers, trading companies, and leisure clubs that established themselves around a bend of the Huangpu
River. These testaments to European colonial power—some of the
tallest buildings in Asia when they were constructed—look as if they
were lifted from the banks of the Thames. They are still there today, a
beautiful and odd part of Shanghai's skyline. Chinese flags flutter atop
every steeple or spire: the modern state's unsubtle reminder that the
colonial era is over.

The French established a concession distinct from the British and
American zones. The area was filled less with grand buildings than
with gardens and residences. Leafy plane trees, common in parks in
London and Paris, lined the streets. Shanghai was the first city in Asia
to adopt modern amenities like public electric lights, a tram line, a
stock exchange, department stores, and cinemas. No wonder it was
then nicknamed the "Paris of the East."

Shanghai was controlled by foreigners, not Chinese, and these foreigners were merchants, not officials. Though a proliferation of sovereignties produced occasional friction, everyone worked harmoniously to make Shanghai a city of indulgence. Well-to-do families could shop New York fashions. Macy's had a department store on the city's main promenade. Those who were in the market for less wholesome fun could find it only a few streets away—at cabarets and jazz clubs, with sing-song girls and Japanese geishas, in Chinese card games and Western slot machines. Shanghai was, in the beginning of the twentieth century, the brothel capital of the world. The city was also full of opium dens, consuming perhaps 90 percent of the world's narcotic drugs. Professional Chinese criminal organizations ran this vice trade and became as powerful as any other political authority in the city.

Shanghai dimmed in the 1930s after Japan began its brutal invasion. Through that decade, the city became a shatter zone of sundry peoples: still the home of Western businessmen, their fortunes made by introducing skin creams, cigarettes, and modern extravagances to Asian buyers; a burgeoning middle class of Chinese who worked in the country's most industrial city; a vast number of itinerant workers, beggars, and orphans who lived in utter poverty; Jews, White Russians, and stateless refugees who lived not much better; and a few ultrawealthy who treated the city as their extraterritorial playground. Leftists organized in these intoxicating settings too. In 1921, a dozen intellectuals gathered in the French Concession to found the Chinese Communist Party.

After Shanghai survived Japan's invasion and Mao's rule, its star rose again through the 1980s. The central government displayed such brazen favoritism toward Shanghai that people waiting for a bus in other cities might call out, "Let comrades from Shanghai board first!" to prompt a burst of sour laughter around them. Today, Shanghai's seedy past is out of view. But remnants of its colonial history are everywhere, only now with a refurbishing by consumer-friendly modernity. Neoclassical buildings made of elegant stone on the west bank of the Huangpu face off against Shanghai's iconic skyscrapers on the other bank, which are once again Asia's tallest buildings, only now encased in glass.

My home in Shanghai was in the former French Concession, which is still full of plane trees and cafés. I loved this area. A twenty-minute walk south of my home was a bakery started by a French émigré, which made apple strudels and baguettes. Twenty minutes north was one of the six Starbucks Roasteries in the world—a two-floor space with a half dozen serving stations—which the company advertises as a "theatrical shrine to coffee passion." Walking twenty minutes east brought me to an attractive gray-brick museum that was the location of the first congress of the Communist Party. Surrounding it is a shopping complex featuring Lululemon, Carhartt, and Le Labo. If any of the summer tour groups found themselves too hot while queuing to enter the Communist Party museum, they could pop next door for a frozen custard at Shake Shack.

When I reminisce about Shanghai, I don't just miss its splendid urban beauty. Nestled throughout these spaces are some of the most wonderful eateries in the world.

Though Sichuan food might be China's most thrilling cuisine, I believe that Shanghai is home to its finest. This region was China's richest and most fertile for centuries, developing sophisticated dishes. Breakfast might consist of a half dozen soup dumplings served in a bamboo steamer, meant to be dipped in a tray of vinegar with a few shreds of ginger. Shanghai noodles are drizzled in scallion oil and served with a slab of braised pork belly and a few pieces of kelp. Shanghai cuisine varies enormously by season, showcasing the bounty of the region. In the autumn, banquet tables are full of steamed mitten crabs, prized not only for their delicate flesh but even more for their bright orange roe, which are briny and have the chewy consistency of the steamed yolk of a duck egg. Spring is even better. Markets lay out a riotous mix of leafy greens, which the Shanghainese like to sauté with a splash of high-proof liquor. Bamboo shoots burst forth when the weather turns warm, and chefs throw them into soups or braises to bring out their sweet tenderness.

Shanghai was wonderful that spring. China's zero-Covid strategy had broadly halted the transmission of the virus. By April 2020, just a few months after the Wuhan outbreak, while Americans were huddling indoors, I was going out again to restaurants and then to cinemas later that summer.* In 2020, when I asked my parents whether I should visit them in Pennsylvania, their reply was not very typical for Chinese: they demanded that I didn't visit. Much better to stay in China, my mom told me, than Trump's America. They were in good shape, and I was glad they didn't need me there.

I didn't dread the virus. I dreaded only the process of reentering China if I departed. One of the core tactics that China used to keep out the virus was to shuttle everyone flying to the country into government-designated quarantine hotels, in which a person would be unable to leave a small room for two or three weeks, depending on the jurisdiction.

So I spent my time inside China, doing things like riding my bike from Guiyang to Chongqing. In 2021, I read giant novels like Dickens's *Bleak House* and Tolstoy's *War and Peace*. I also met my now-wife Silvia, a professor at the University of Michigan who was taking a sabbatical at NYU-Shanghai. As an ethnographer of technology cultures, Silvia had lived in China and continued to stay engaged. The United States that Silvia departed in 2021 was still a distressed place, where few people were getting together for in-person contact. She was even more thrilled than usual to be able to return to China for

---

* Tragically, the only movie that was available to watch then was Christopher Nolan's baffling *Tenet*, which might have been better than no movie at all.

research when she obtained a rare visa. After completing her quarantine, Silvia felt a sense of freedom on Shanghai's vibrant streets. We got to know each other as we cycled around the city to cafés and dumpling shops.

But things weren't completely normal. To enter most public spaces—my office, a restaurant, even many outdoor commercial areas—I had to pull out my phone to display my contact-tracing QR code to the burly man guarding the entrance. Green meant normal, while yellow meant that I had had some degree of proximity to a positive case; one wouldn't have needed to flash a red code, since the state would probably have hauled that person off to quarantine. The cell towers that triangulated a person's location and the contact-tracing workforce sometimes produced errors. Merely walking by a restaurant with a known infection might turn your code yellow, even if you never went in. People often complained that no would explain why their code stopped being green. But quarantines and movement restrictions felt like inconveniences worth respecting. Having to fumble for my phone to pull up my contact-tracing app whenever I entered a public space didn't feel like too big a deal when I looked at how other nations were suffering. China was piling on these controls gradually, so the incremental asks felt more acceptable.

But I was aware of the ground shifting under my feet. A big part of my work was to cover US–China relations, which had been unraveling even before the pandemic. Throughout 2020, President Trump lobbed shots at China's tech companies, which I covered even while travel became more difficult. But what was happening inside China was even more unexpected. Xi Jinping grew bold as China controlled the virus and the rest of the world did not. While he announced a campaign to achieve "common prosperity," he cracked down on digital platforms and real estate developers. My clients had a lot of questions for me since there were few people they could call who were in China. They asked me how it was possible that after the virus emerged in Wuhan, China seemed to be containing it better than anyone else. I answered them honestly: China was doing well—for now.

In December 2021, we began hearing about the omicron strain of the virus, which scientists told us was so much more transmissible than earlier variants. I wondered about omicron's effect on China in my annual letter published on the last day of that year. "I worry that it's so transmissible that the government will . . . implement lockdowns far more severe than anything it has done to this point." On Twitter, I was more flip: "I prepared three items at home to survive a potentially severe lockdown: mooncakes (high-caloric and long-storing); a bike with a trainer (to cycle through the metaverse); and the Hebrew Bible (Robert Alter translation)."

Xi'an had already given us a preview of what it took to control a more transmissible strain of the virus. At the end of 2021, the northwestern city entered a lockdown. Residents ran out of food in the middle of winter, and horrific stories started to emerge. A woman who was eight months pregnant felt pains and wanted to be admitted to a hospital but was refused entry by staff until she took a PCR test—which could take several hours to process—and provide a negative result. Two hours later, she started bleeding heavily. She miscarried outside the hospital while pleading to be let in. Her story went viral until censors deleted it.

The pleasures of Shanghai curdled in the spring of 2022. Few people were able to buy seasonal greens or bamboo shoots. The central government had ordered a lockdown for the city of twenty-five million, who were mostly unable to step foot outside their residence for two months. For most of the pandemic, Shanghai distinguished itself by confronting the virus with a light touch. It might have counted as a triumph of the engineering state. Then Shanghai suffered what was probably the most ambitious quarantine that any state has ever attempted.

Omicron descended on Shanghai as spring commenced. At the

start of March 2022, city officials announced that a quarantine hotel holding people flying in from overseas had bungled its safety protocols, leading to the infection of a few cleaning staff, who brought it into their communities. Shanghai implemented its now-familiar pandemic playbook: Authorities conducted mass tests, brought people who tested positive into a centralized quarantine facility (usually a sports stadium or convention center with thousands of beds), tracked the location history of each confirmed case, and imposed a lockdown on the neighborhoods where close contacts lived.

The premise of a lockdown was simple: nobody would be allowed to leave their apartments except to have their noses or throats swabbed in a government-administered PCR test. Nearly everyone in Shanghai lives in apartment compounds made up of several buildings with a courtyard below. The building I lived in was a smaller walk-up with six floors that held a couple dozen households. Most people, however, live in taller developments, which could each hold a few hundred households. High-rises might be more desirable in normal times, but I would soon find out how lucky I was to live in a walk-up. Huge developments were exponentially more likely to suffer a lockdown, as a single case could condemn the whole building.

Shanghai's pandemic playbook had halted prior outbreaks. This time it failed. Throughout March, the number of new cases rose each day. Seven-foot-tall plastic barricades sprouted up around apartment compounds throughout the city, signifying a positive case. Restaurants, cafés, and other businesses shuttered. As commercial sounds dimmed, voices from loudspeakers became a constant presence. Several times, they summoned everyone in my compound to go to a nearby facility so that we could all be tested. I never tested positive, thankfully. Nobody I knew did either. If you were positive, the government would take you into a mass quarantine facility; if health authorities suspected you had the virus based on your location history, you might be prohibited from leaving home for a few days. Several of my friends were told they had to stay inside because they were proximate to someone who might have the virus.

By late March, a sense of dread pervaded Shanghai. On a particularly eerie day, Silvia and I heard from three separate friends within the span of an hour that they were no longer able to leave home for three days: A neighbor had been a close contact to a positive case. That morning, Silvia and I cycled to a café near the Embankment Building, an iconic art deco residence that once housed Jewish refugees. We remarked over some croissants that the city had never felt so quiet. As we cycled back home, we saw the Embankment Building transformed. Health workers and police officers had covered the exits and were helping each other put on all-white protective suits held together with blue tape. They looked like they were preparing to lay siege to the building. These workers, nicknamed *dabai*, or big whites, became dreaded specters that symbolized enforcement of zero-Covid.

In the government's daily press conferences, Shanghai officials repeatedly denied that they would order a broad lockdown for the city. The situation was in hand, they told us, even though the number of new infections was increasing every day. "Shanghai Has No Plans for City Lockdown" read a headline on March 24 in the state-run *China Daily* newspaper. Shanghai is "too important to lock down," claimed Wu Fan, one of the members of the city's health commission, during a press conference on March 26. Then she added, with a shade of arrogance, "The city of Shanghai does not belong only to the people of Shanghai. It is a driving force for the global economy, and a lockdown here would shake the world."

The day after Wu Fan's defiant proclamation, Shanghai announced it would lock down. The announcement was ever so softly worded. Shanghai was enacting a "partial pause" to enter a "quiet period" that would last eight days. First the eastern half of the city would enter lockdown, then the western half. The city ordered people to work from home; all businesses would shut down. The bridges and tunnels connecting the two halves of the city (separated by the Huangpu River) were blocked. The government promised to deliver food and ensure medical access. All lockdown measures, they said, would terminate on April 5.

Shanghai's lockdown extended far beyond that date. Case counts exploded while the city was in its quiet period. Instead of lasting eight days, the lockdown lasted eight weeks, finally reopening in June. I often think about the *China Daily* headline "Shanghai Has No Plans for City Lockdown." It could be read in two ways. I first understood it as a denial that the city would impose a lockdown. I understand it now as a totally accurate explanation of what happened next: The city had made no plans for confining twenty-five million people to their homes for eight weeks.

Government drones descended throughout the city. Since the start of the pandemic, the state had dispatched megaphone-equipped drones to nag the uncompliant. A person walking without a mask might hear a whirring craft above his head, from which a distorted, barking voice would yell at him to mask up or return home. A Shanghai neighborhood official outlined what would happen if a drone came upon an illegal gathering of people: "The drone will try to dissuade," in other words, berate them, "and ground forces will be linked in real time."

An even more bewildering use of drones took place in the early days of the Shanghai lockdown. The city's top mental health official introduced an unexpectedly sparky phrase in an otherwise drab press conference on the course of the virus, demanding that Shanghainese "repress your soul's yearning for freedom." Social media users immediately began to make fun of the phrase by putting it into memes. People weren't used to poetry from bureaucrats. One night in April, as the lockdown swung into high gear, a drone carrying a megaphone began blasting that message into apartments full of huddling residents: "Repress your soul's yearning for freedom," with a woman's voice played on loop while a light blinked from the drone. "Do not open your windows to sing, which can spread the virus."

The phrase stopped being amusing.

. . .

OVER APRIL 2022, STRESS in Shanghai spiked to unimaginable levels.

The primary worry for most people was how to secure food when they could not leave their homes. The surprise lockdown announcement, coming in the evening, gave people in Pudong, in the eastern half of the city, only hours to stock up on food. Puxi, the more populous western half where I lived, had four more days to prepare. Many people had failed to stockpile essential goods, after repeated denials of lockdown by city officials diminished their sense of urgency. Even among people who were able to stock up, it was difficult to keep fruit and vegetables fresh after ten days or so.

The Shanghai government had promised to make food deliveries. It started out okay: Everyone I know in Shanghai received a handful of packages featuring a welcome but random assortment of fruits, vegetables, and meat. But government deliveries quickly ran out of steam. On April 5, when the lockdown was supposed to end, Shanghai announced that it would need to be extended. That's when food concerns heightened. By mid-April, nearly all of my friends had experienced at least a few days of food insecurity. Two sets of parents told me that they forfeited their own meals to save food for their young children. When Emma, an American friend of mine, opened her government-organized food delivery, she discovered a freshly slaughtered chicken, still with a few feathers on it. She had no idea how to prepare it; she also had nothing else to eat. So she went to YouTube. After psyching herself up, she pulled up a video to learn how to gut a chicken, grimacing as she eviscerated it.

Without help from the government, people tried to place orders on grocery delivery platforms. They became immediately overwhelmed.

The thing to do was to set a lot of alarms—6:00 a.m. for Meituan, 6:30 a.m. for DingDong, 7:00 a.m. for Freshippo, 8:00 a.m. for Yonghui—to place an order in the half minute before all the food was snapped up on these platforms. The food supply chain broke down for several reasons. One of them was that the state made it difficult for truckers to bring food into Shanghai, fearing that they could bring the virus across vast distances. To cross a province, truckers often had to wait in lines, remaining in their cabs until their Covid test results were available. One viral video showed a driver holding up bottles of his own excrement because traffic control would not permit him to exit. He exploded in frustration that the controls made him feel like an "animal in captivity." These strictures drove many to quit. In mid-April, trucking activity in Shanghai was only 15 percent of its normal level.

Much of the food that made it into cities rotted before it could be delivered to residents. The responsibility for organizing food deliveries fell to Shanghai's neighborhood committees, the lowest level of officialdom, which had been staffed mostly with elderly volunteers more used to propaganda work than the intensive engagement with Excel spreadsheets that the logistics of food parcels demanded. The state also immobilized normally robust food courier services. Delivery workers wearing mustard yellow or baby blue uniforms, carrying food inside a box strapped to the backs of their scooters, faced lockdowns too. A few made the choice to be homeless in order to continue work. At the cost of sleeping under bridges or in other public spaces, they were able to roam around the city, delivering food to earn higher commissions.

Shanghainese marveled that they could worry about hunger while they lived in China's richest city in the year 2022. People muttered darkly that China had achieved "common prosperity," Xi Jinping's new signature initiative to reduce inequality, in China's most capitalist city a decade ahead of schedule. Though some people connected to the government might have had better access to food, nearly everyone—

rich and poor, young and old, local and foreign—was in the same hungry boat. Celebrities complained online that they had to spend nearly $300 to have some vegetables and eggs delivered. One of the country's top venture capitalists, who was an investor in grocery delivery companies, sent a message on social media asking people how to get food.

After complaints about hunger grew louder, the Communist Party responded with a time-tested tactic: scapegoating. State media publicized a few cases in which food deliveries were hoarded by residents rather than being distributed to their neighbors. These cases might have been real, but they weren't the main problem. The fundamental issue was that the surprise lockdown announcement had deeply broken Shanghai's food supply chain, crippling both long-distance and local deliveries.

In the latter half of April, people found a lifeline. My friend Owen had moved to Shanghai from Beijing just a few months before the start of the lockdown. An American in his early thirties, Owen went to work at a policy research outfit in Beijing after graduating from college. He lived on his own in a modest walk-up, above a noodle shop, that looked out toward a small supermarket. Since Owen lived in Puxi, he had more time to stock up before he was locked down. The day after the announcement, he woke up early to go to the grocery store, finding a long line even before doors opened. He managed to snatch a few bags of fresh vegetables along with ground beef. These he cooked into a bolognese sauce, which he kept frozen in several parcels.

Shortly after the lockdown began, Owen received a generous bag of government-sent rations: peppers, tomatoes, bok choy, garlic, ginger, potatoes, and more. A smaller bag arrived the following week. Then nothing. For weeks, the government dropped off no new food. Owen began trying each day to book a grocery delivery but never succeeded. Everyone else in the city was trying to do the same, fighting for a small pool of available food. After a few days without success, he thought to himself, "This could be bad."

Owen wrote his WeChat handle onto a slip of paper and taped it outside his door. WeChat is the universal chat app in China, and a typical person belongs to several dozen chat groups: family, colleagues, other school parents, board game enthusiasts, friends from college, any group with activities to coordinate. Since Owen lived on the lowest residential floor, just above the shops, everyone was able to see his WeChat handle when they passed by to take Covid tests. Soon enough, the whole building added him, and he formed a groupchat for its thirty-six households. Owen didn't seek to be his building's unofficial representative. As a tall dude with blue eyes who had just moved to Shanghai, he wasn't the likeliest spokesperson for his all-Chinese building. "After my neighbors added me," Owen told me, "their attitude was 'what's the plan, bro'?" He became the point person for communicating with authorities as well as the ringleader for organizing communal functions.

Neighbors were able to coordinate help for each other in this WeChat group, even elderly ones (though they might be digitally represented by a son or daughter who wasn't living with them). The chat's most important function was to arrange group buying, which Owen accomplished by placing bulk orders directly from a wholesaler. Somehow, food deliveries were possible that way. The system eased hunger through the latter half of April, though it still had a lot of problems. Bulk orders demanded averaging out food preferences; everyone wanted eggs, but not all foreigners could convince their Chinese neighbors that butter was necessary too. One day Owen found himself craving good bread, a luxury that his neighbors wouldn't have agreed to. He bought some from a home baker across town, at $40 a loaf.

Once food arrived downstairs, a rotating cast of volunteers delivered it throughout the building. Some walk-ups agreed to prohibit, for example, the purchase of plastic jugs of water, since it was unfair for neighbors to lug them up flights of stairs. Owen only sometimes volunteered for these jobs, since he still had a day job to do at a public affairs company. My friends felt they had to do two full-time jobs: their regular one and the many hours a day spent trying to procure food necessities. Bulk orders weren't possible for everyone. Smaller buildings didn't have enough residents to place egg orders by the thousands. And the system disfavored the elderly, who struggled to navigate mobile purchases.

While Shanghai was in strict lockdown through April, the number of new infections kept rising. The lockdown extension surprised no one. Everyone knew that lockdowns wouldn't end until numbers dropped to zero.

I asked Owen why so many people still caught the virus during lockdown. "It was for sure due to the tests," he replied. People had to report for Covid testing nearly every day, sometimes twice a day. A medical team would enter an apartment compound and summon everyone downstairs, either on WeChat or with a bullhorn. Anyone who didn't come down would receive a buzz on the downstairs gate; if that didn't work, they would hear a knock on their door. It was absurd that elderly people—some of whom rarely left their apartments without a pandemic—were squeezed into elevators with neighbors.

It's impossible for anyone to be certain how exactly they caught the virus. Perhaps omicron was so transmissible that people caught it through the plumbing or ventilation systems that connect people in Shanghai's apartments. Perhaps it spread through food deliveries. Most people believe they caught it through the daily testing regime: from a neighbor while they were waiting in line. Every so often, a story popped up that the medical worker swabbing everyone's throats had the virus himself, which at least contaminated your sample and perhaps infected you. Despite exacting measures, the number of new confirmed cases kept rising for four weeks until the end of the lockdown.

Many people feared the virus itself: For two years, the Chinese government did everything it could to frighten people about getting

Covid. Censors stepped in to make sure that no one called it "just a cold." If one tested positive for the virus, life became a lot more complicated. The Chinese government did not permit people who tested positive to stay in their homes. Since the early days of the Wuhan outbreak in 2020, authorities realized that someone who had the virus inevitably gave it to their entire household and perhaps the entire building. Health authorities came to take the infected away to one of the huge, centralized quarantine facilities. It wasn't fun to be in these places. A producer from CNN who tested positive for the virus described the unpleasantness of living in Shanghai's largest convention center, which hosted fifty thousand beds. She described lights that never turned off, loudspeakers demanding that everyone show up for PCR tests at 6:00 a.m., and everywhere the stench of toilets or unwashed laundry.

After taking the infected to quarantine facilities, health authorities entered people's homes to sanitize them. That meant dousing everything in disinfectant—furniture, books, electronics, clothing, the piano. Pet owners faced a particular dilemma. They might ask a neighbor to look after a cat or dog while they were away in quarantine. Those who couldn't find help decided, painfully, to release their pet into the streets and hope for the best. It was that or leave it indoors, somehow providing enough food to sustain it through the uncertain length of the owner's quarantine period. A viral video of a *dabai* chasing down a corgi with a shovel, striking it until it lay prone, did not make the decision any easier.

Parents of young children were even more frightened. Shanghai practiced a policy of separating babies and infants from parents, even if both tested positive. Photos spread of infants crying as they were held in metal cribs, while panicked parents told media that they hadn't received updates from hospital staff on the status of their children for days. One woman told a reporter that the virus no longer frightens her. "Separation from my loved ones scares me more than anything else." After an outcry online, the city dropped its policy of isolating children.

One day, Owen felt a slight pain below his abdomen. When he looked down, he saw a bump the size of a small fist between his right thigh and his groin. Googling suggested that it was a hernia: Part of Owen's small intestine popped out and couldn't be tucked back in. It's not an uncommon problem for men, though usually not until they're older. The good thing was that the bump didn't cause too much pain. He's still unsure how it developed. Possibly, he told me, from a strong sneeze, compounded by all the stress while being overwhelmingly sedentary.

Owen decided not to seek medical attention. It was nearly impossible to get to a hospital. One of the stories that provoked wide outrage was the case of an asthmatic forty-nine-year-old nurse in Shanghai, who was denied treatment at the hospital where she worked before she collapsed and died. People with health conditions were gripped by fear that their medications would run out: Attempting to procure them might have constituted yet another full-time job. One of my colleagues told me that her uncle with diabetes died during Shanghai's lockdown because he could not access dialysis treatment. People marveled that hospitals more or less ignored every medical condition aside from Covid infections.

There was no universal experience of the Shanghai lockdown. The city of twenty-five million people dealt with situations ranging from the nightmarish to the merely difficult. Not everyone experienced hunger: Certain compounds found fairly regular access to food, especially if a government official lived in the building. Introverts found ways to create structure in their lives. After the lockdown, people got to know the neighbors they had previously interacted with only on WeChat. Even for those who found it all bearable, the challenge was that no one knew how long the lockdown might last. Neighborhood officials grew uncommunicative, mostly because they had little idea of when the lockdown would end. Perhaps the most unnerving features of the pandemic were the frequently changing government policies. People had little idea when they would be able to go outside for anything other than lining up for a Covid test, while they spent exorbitantly on food chosen by their neighbors and tried to stay sane and healthy.

For many, there was nothing to do but stay glued to their phones all day, idly scrolling through entertainment or frantically attempting to secure a grocery delivery. Or they spent time on social media. A lot of what we know from Shanghai's lockdown comes from the videos shared on WeChat, Weibo, and other platforms.

The bulk of Shanghai's population experienced at least a few moments of immense frustration. Banging pots and pans during the night became a much-shared form of protest. A few videos portrayed whole buildings of people engaging in cathartic screaming (which might explain why the government sent drones instructing people to stop "singing"). Someone shot a video of a woman wandering stark naked around her courtyard. Many videos purported to show the aftermath of people who committed suicide by leaping from their high-rise to the ground below. People shared videos of others screaming denunciations of the police or the regime. A couple who had tested negative for the virus filmed themselves confronting a police officer who insisted on taking them to a quarantine facility. When they showed him their negative test results, he replied, "You are positive if I say you're positive."

China's already formidable censorship regime distinguished itself in this crisis, meeting the challenge with a staggering response. Complaints and protest videos were deleted quickly after they went viral. When Shanghai residents posted en masse the first line of China's national anthem, "Arise, you who refuse to be slaves," their posts were removed. Censors took down posts spreading a National People's Congress spokesperson's remark that quarantines may be unlawful.

At one point, social media platforms blocked the word "Shanghai" from search results.

One video managed to achieve censorship escape velocity. Someone (or a group of people) collated a chronological montage of audio clips into a video titled "Voices of April." The six-minute clip included Wu Fan's remark that Shanghai was too important to lock down; shouts of people demanding food; a man pleading for his sick father to have medical treatment; exhausted officials saying there was nothing they could do. "Voices of April" dominated my WeChat feed for a few days. People put more effort into sharing that video than anything else in an attempt to circumvent censorship. They even put it on the blockchain, where it will remain for posterity.

By late April, most of my foreign friends—especially those with children—departed China, a few for good. Pricey plane tickets were the least of their concerns. To depart from their apartments and get to the airport, people had to sign affidavits swearing not to return to their residence. A taxi to the airport that costs $30 in normal times shot up to $300 because only a few cars and buses were permitted to pick up passengers.

The number of new infections in Shanghai peaked in late April. Food logistics improved through May, such that Morgan Stanley was able to do what an American bank does: deliver extravagant gifts to select clients. One of my friends received such a package and told me that it included a crayfish salad, which felt like an absurd luxury at that moment. On June 1, the government gingerly allowed the city to return to normal.

•   •   •

SHANGHAI'S LOCKDOWN WAS ONE of the major turns in the dramatic arc of China's pandemic experience. Over the three years of the pandemic, the emotional lives of people across the country veered from fury to pride to desperation.

The first act took place over the early days of 2020. I was living in Beijing and watched as the city descended into anxiety as we heard about the coronavirus that emerged from Wuhan. By the start of February, Beijing's streets were empty while Wuhan's were positively grim. We were galled by the death of Li Wenliang, a doctor in Wuhan who faced police reprimands when he attempted to warn people of a new respiratory virus. On February 7, he died from the coronavirus that he attempted to warn people about. That night, my WeChat feed was dominated by tributes to Doctor Li, accompanied by immense fury at how the police had treated him. I wouldn't see my WeChat feed be so dominated by a single event until "Voices of April" two years later.

Officials in Wuhan suppressed news about the new virus circulating in their city for the most picayune of reasons: They wanted to ensure the smooth operation of an annual political meeting. In those crucial early days of the pandemic, they wanted to hear no news that anything was amiss, especially not since the Lunar New Year was about to begin. Wuhan officials refused to call off a community feast that attracted a hundred thousand people only six miles away from the Huanan Seafood Market, where the coronavirus was already circulating. At the Lunar New Year gala, state media praised performers for helping the show go on even while they were sick.

Beijing felt grayer and colder in February 2020. Nearly all restaurants and public spaces were closed. My friends and I went out on bike rides across mostly empty streets. Meanwhile, frightening images were spilling out of Wuhan. The official narrative I heard in Beijing was of heroic sacrifice. Some of the images that state media released were inspiring: Authorities livestreamed a dozen excavators that built a new hospital in eleven days. But videos of nurses crying while having their heads shaved (to prevent virus transmission) did not make good propaganda. The unofficial narratives were far more heartbreaking. One forty-two-year-old woman living not far from Beijing produced one-line snapshots of personal stories that she posted on social media:

> *The one following a hearse in the deep of night, calling out "Mama" in grief.*
> *The 12-year-old who went alone to report their orphan status after their entire family died.*
> *The one who was forced to write "You must wear a mask when you leave the house" 100 times by the local police.*
> *The one who carried their mother on their back while searching everywhere for treatment, walking for three hours.*
> *The one who recovered from a severe case only to come home to find their entire family dead, who hung themselves from the roof.*

Then she stopped posting. Police in her hometown published a notice a few months later that she was guilty of spreading rumors and sentenced her to six months in jail.

Xi Jinping declared controlling the coronavirus to be a people's war, a Maoist term that promised to smash imperialist invaders with guerrilla maneuvers. The state marshaled hefty men, *dabai*, dressing them in ill-fitting white uniforms and arming them with temperature scanners to check whether people entering buildings had a fever. Crimson propaganda banners that previously declared the superiority of socialism were replaced by ones urging people to stay indoors. The government pulled out all the stops to prevent people from traveling across the country. It halted train services, preventing the millions of migrant workers who traveled home to celebrate Lunar New Year from returning to their workplace. And it blocked nearly all international flights. The trickle of people entering the country were mostly Chinese nationals who could accept living for up to three weeks in quarantine hotels.

I felt baffled and angry that winter. Covid-19 was China's third epidemic in three decades, exploding in exactly the same pattern as the previous two. In the 1990s, Henan province suffered an AIDS outbreak after blood banks reused needles and commingled diseased blood with healthy blood; the government spent years silencing whistleblowers in this slower-moving epidemic before finally confronting the problem. In 2003, officials in Beijing and Guangdong attempted to suppress news of the SARS outbreak before moving decisively to control it.

A year before the coronavirus spread from Wuhan, China's top disease control official, George Gao, offered a boast: "I am very confident to say that SARS-like outbreaks will not occur again because the infectious disease surveillance system network of our country is well established." Gao had it right when he said that China had developed a technically impressive disease surveillance system. What he failed to factor in were the weaknesses in China's political system, in which local officials prevented health workers from reporting the disease. Rather, Wuhan officials directed police to punish medical whistleblowers. And so China faced its worst public health crisis yet.

The second act of the pandemic began in March 2020. The authoritarian impulse to suppress bad news produced a catastrophe; then the restrictions that the state imposed on daily life beat back the virus. While life was starting to return to normal for those of us in Beijing, Covid-19 had slammed into the rest of the world. And though none of us forgot our anger with how it all started, people inside China watched as much wealthier governments bungled their own pandemic response. Local officials in Wuhan and Hubei province weren't the only ones who denied the seriousness of the virus; few global leaders took it seriously. While the engineering state activated every sinew of its powers to break the transmission of the virus, most other governments were treating Covid almost as if it were a curiosity that could affect only Chinese people.

In my professional life, I was puzzled that even financial markets barely reacted to major lockdowns in China. In February, I went on Bloomberg's *Odd Lots* podcast, talking with cohosts Tracy Alloway and Joe Weisenthal, as we all felt a bit bewildered that the market didn't want to price in a global pandemic. Either these hyperrational people thought that Beijing's measures were effective enough to stomp out the virus or that it couldn't really affect the rest of the world. Reality settled in shortly afterward.

The fierce anger that the Chinese people felt at the cover-up in Wuhan partly transfigured into a sense of pride at the pandemic control efforts conducted by the central government. Chinese saw how Italy, Russia, and the United States mishandled their pandemic response. They gawked at clips of Donald Trump speculating that the virus would disappear on its own or that it could be tackled by injecting disinfectants into people. When Li Wenliang died, foreign commentators bandied about the term "Chernobyl moment" to describe the greatest threat to the Communist Party's legitimacy in decades. Three months later, Xi declared that China had "turned the tide on the virus." Subsequently, while the miseries of Covid deepened in other countries, *People's Daily* declared that pandemic controls were a demonstration of the superiority of China's socialist political system.

China's pandemic control measures were not unique. Japan, South Korea, and Taiwan also imposed lockdowns, practiced centralized quarantines, forced international travelers into quarantine hotels, and demanded everyone display health tracking apps. But China enforced these controls more diligently, on far more people, because it is an engineering state.

Only a country ruled by engineers could be so single-minded about pursuing a number. Since the early days of the pandemic, Chinese officials became obsessed with two numbers: new infections and the reproductive rate of the virus. The engineering state did everything it could to stomp them down. It led, ultimately, to the pursuit of zero-Covid (formally known as dynamic zero clearing in Chinese).

Just as with the one-child policy, the target could not be clearer: The number was in the name. And just as with the one-child policy, zero-Covid was suffused with military language: China was fighting a people's war against the virus, and cities like Wuhan and Shanghai were battlegrounds that had to be won.

No policy was too senseless to pursue after Xi Jinping staked the prestige of the Communist Party on control of the virus. "Prevention and control work cannot be relaxed," Xi repeatedly instructed local officials. The costs of zero-Covid seemed worth it for a while. Only later did the increasingly severe movement controls and state disregard for any medical condition except Covid turn the strategy into a farce. Officials brought a literal-mindedness to enforcing zero-Covid that created situations best described as whimsical. The coastal city of Xiamen swabbed the mouths of fresh-caught fish to test for Covid. A panda research base in Chengdu tested every animal in its facility. Medical workers chased down Tibetan and Mongolian herdsmen—who probably saw nothing but yaks for days on grassland steppes—to swab their mouths.

Throughout the three years of the pandemic, China developed a weightier state apparatus, one better able to impress itself upon its subjects using digital surveillance. Enforcement of the one-child policy was an intensely physical act, in which health workers got up close and personal with vulnerable women. To achieve zero-Covid, the state once more mobilized millions of people: a mostly male workforce that donned white protective gear to become *dabai*, or the public enforcers of pandemic control, and a mostly female workforce that worked as contact tracers to investigate people's travel histories between and within cities.

Digital technologies gave the engineering state a tool it did not have when enforcing the one-child policy. Implementing zero-Covid was a technologically intensive affair that used mobile networks to track people's movements, sometimes aided by facial recognition technologies and other forms of digital surveillance enabled by the mobile devices that nearly everyone carried.

Sometimes, China's digital platforms introduced helpful interfaces, for example, when mapping services made it easier for people to find fever clinics nearby. Sometimes, they were enlisted to control the movements of people. To access the showers at Shanghai University, students had to display a code on their phones, which was green for five and a half hours every two days. A sociology student marveled at her experience to a Shanghai newspaper: "It's such a strange feeling: the idea that all our daily activities—what we eat, or when we can take a shower—are included in the authorities' plan." The state attempted to reduce movement throughout society. Since Chinese university campuses were already self-enclosed areas, often far from urban zones— and since college students are meant to spend all their time studying anyway—officials simply decided to lock them in. During lockdowns, students struggled to stay sane in their dorm rooms, which might have four people bunking together.

Factory workers were sometimes enclosed within a "bubble." That was Beijing's invention for the 2022 Winter Olympics, in which foreign athletes were physically separated from the rest of the population. Companies attempted to create bubbles by enticing workers never to leave the factory—sleeping by the assembly lines— for perhaps quadruple their usual pay. Volkswagen and Foxconn, for example, adopted these bubbles to keep the assembly lines for their cars and iPhones flowing. The problem was that even the most persevering workers grew tired of living at the assembly line after a few weeks. And as often as not, the virus would still penetrate the bubble, infecting everyone.

Public spaces sometimes suffered a surprise lockdown. On more than one occasion, visitors to Shanghai Disneyland were told that they could no longer depart from the happiest place on earth because a close contact of a confirmed case had passed through it. Thirty thousand visitors were trapped inside the park for much of a day in 2022, departing only after they all tested negative for the virus. There didn't seem to be too much grumbling since Disneyland continued to operate its rides. Better there than at Shanghai's Jiuting Bridge Wholesale Market or the Songjiang Building Supplies Market, both of which kept more than a thousand people locked up for days without providing water or food. Throughout the latter phases of zero-Covid, panicked white-collar workers streamed out of office buildings in Shanghai or Shenzhen when there was a rumor that a building might be placed under lockdown. It's not clear what was more frightening: being trapped with coworkers or not being able to shower.

Big cities attempted to enforce lockdowns on targeted spaces, like a particular apartment or office building. Covid lockdowns were far more indiscriminate elsewhere, for an entire city might be locked down over the discovery of a handful of cases. Smaller cities had little confidence that the medical infrastructure could handle a surge of infections, so local officials were quicker to order disruptive lockdowns. The people who lived in China's border cities (next to Myanmar and Laos in the south, or Russia and North Korea in the north) were subject to the most frequent lockdowns of all, since people traveled through sometimes porous national borders.

Many people made their peace with these practices because they listened to health authorities who said that it was better than infections and deaths; because the state piled the regulations on gradually, improvising as it went along; or because they had no other choice. By the time that Xi'an and Shanghai's lockdowns came into view, however, more people questioned whether food insecurity and indefinite confinement made the pursuit of zero-Covid still worthwhile.

The first act of zero-Covid was characterized by fury, the second act by pride mixed in with some degree of exhaustion. The third act, which began after Shanghai's lockdown, led to desperation and, later, broad protests.

. . .

AN HOUR AFTER Shanghai's surprise lockdown announcement, Silvia and I purchased airline tickets to Yunnan, the mountainous province in China's southwest where my family is from.

Neither of us trusted that the Shanghai lockdown would last only eight days. More important, both of us were able to work remotely. We had been discussing whether to depart from Shanghai since the eerie day we saw the *dabai* besiege the Embankment Building. The lockdown announcement was a good prompt for organizing our departure. When Silvia and I left, our flight was one of a dozen that hadn't been canceled that day. We were lucky. Cities across China had already refused to allow flights from Shanghai because it was the center of the omicron outbreak.

A trip we thought would take two weeks turned into something that lasted nearly half a year. Yunnan is a good place for reflection because it is plausibly China's freest province. In the mountains of Yunnan, I glimpsed the idea of not only the engineering state but also the lawyerly society. Several questions ran through my mind while I was unable to return to Shanghai: How was China able to enforce lockdowns of this scale? Why have people been able to accept it? And when will Xi finally give up these controls?

Yunnan is even more mountainous than neighboring Guizhou, and still less touched by the industrial transformation of China's prosperous coastal zones. It remains one of China's poorer regions, its economy sustained by tourism and resource extraction, particularly minerals and tobacco. The northernmost part of Yunnan is historic Tibet, home to a chunk of the Himalayas—including Kawa Karpo, one of the most sacred mountains in Tibetan Buddhism. Shangri-La is the largest city in the region. The small roads around Tibetan monasteries are strewn with prayer flags and studded with impassive yaks. In Yunnan's south, where I flew from Shanghai, the mountains are greener and gentler. Tea hills and rubber plantations rise above the Mekong River, carrying snowmelt from Tibetan highlands that eventually drains beyond southern Vietnam. Xishuangbanna is one of China's most biodiverse regions, home to many trees and plants, wild elephants, peafowl, bears, and birds galore.

Around half of China's officially recognized minority groups have their homes in Yunnan. They live between its snowy mountains, rainforests, rice terraces, and fast-moving rivers. Many of them have historically resisted rule by the dominant Han. It is part of a vast zone of highland Southeast Asia that various scholars have labeled Zomia, which holds innumerable hill peoples who have developed state-repellent practices. James C. Scott has written most elegantly about how people in Zomia have become "barbarians by design," who cultivate shifting root crops (which are less assessable by tax collectors) and maintain an oral culture (which makes their histories and ethnic identities more malleable). It is not surprising that people in these hills claim various liberties, like foraging wild mushrooms, hunting game, or trafficking harder drugs. Not even the Han-Chinese state has been able to assert jurisdiction over the dense jungles and rugged mountains of the region.

Silvia and I spent several months in the city of Dali, which has a lake on one side and a mountain range on the other. The local Bai people built lovely lakeside houses, made of white walls ornamented with wooden carvings and blue ink paintings. The Bai are mountain farmers who have a long culture of craftmaking, producing marbleware or indigo linens for trade with the Han. Up until the 2000s, a different Bai product attracted foreign travelers: cannabis, which grew freely in the region. Foreigners in Beijing or Shanghai may still reminisce about the good old days in Dali, where one could be beckoned by a smiling older lady into an alley to purchase a baggie.

With its lake, nature, and sunny weather, the city has gained the nickname of Dalifornia. While I continued my work remotely, Silvia was doing ethnographic fieldwork. She introduced me to some of the young people there, who were exploring their interests in agriculture or virtual technologies. The city has attracted an odd mix of people: China's burgeoning organic movement, which is mostly made up of younger people who want to take advantage of Dali's fertile soils; moms who bring their kids to experience nature-focused educational programs as a break from the hypercompetitive schools in Shanghai and Shenzhen; and foreigners who came for the cannabis and stayed for the slower pace of life, opening sourdough bakeries, cafés, and techno clubs. Nowadays young Chinese and foreigners go to Dali not for cannabis but for more thrilling drugs: cryptocurrencies, NFTs, and other Web3 paraphernalia.

A great deal of China's crypto community has relocated to Dalifornia, drawn as much by the beautiful natural setting as the permissive environment. Mountains have beckoned, as Scott has written, to dissenters, rebels, and subversive types. It is not only the air that thins out at higher altitudes: The tendrils of the state do too. Small bands of people tired of tax administration or the other ills of governed life have climbed upward. As a consequence, mountain dwellers tend to be seen as unruly folks, be they Appalachian Americans, Highland Scots, or various ethnic groups in Yunnan and other parts of Zomia. What is a difficulty for government administration and industrial growth is often a positive for personal liberties. The mountains of Yunnan protected local peoples from the state-produced famines in the Great Leap Forward and the harangues of the Red Guards during the Cultural Revolution.

That's why Yunnan might be China's freest region. It is farther from the country's core, and unlike in Xinjiang or Tibet, the state hasn't treated its ethnic populations to its most stringent controls. Yunnan can be a hub for drug trafficking, cryptocurrency gatherings, or the most radical activity in recent years: lax Covid enforcement. Local governments closed a market here or there throughout the years of zero-Covid but didn't bother to enforce the serious lock-downs that affected Wuhan, Xi'an, and Shanghai. Too few people lived in enclosed apartment compounds for that to work. If author-ities squeezed too tightly, people in Bai villages might have simply walked from their backyards into the mountains.

I picked up the idea of the engineering state in Yunnan's moun-tains. The government was able to treat people as chess pieces to move around (or hold still) in Shanghai, while failing to do so in more remote areas. A glimmer of the lawyerly society came into view as well. One of the most-shared essays during the lockdown was a com-mentary by Tong Zhiwei, a constitutional law professor in Shang-hai, who pointed out that the city's lockdowns had no legal basis. The government's response to Tong's legal arguments was to censor his essay and erase his social media profile. What did it matter that keeping twenty-five million people indoors over an undefined period lacked legal authority? Good luck to anyone attempting to go to a courthouse to file a suit.

It is a little bit difficult to praise the US response to Covid. In ret-rospect, the whole thing looks shambolic, with different states having different policies, mostly made worse under Donald Trump's chaotic management. Americans stumbled into learning to live with the virus in large part because of the ineffectualness of the government. But the United States (under Trump's Operation Warp Speed) produced mRNA vaccines that China could not. And in retrospect, China's response to Covid looks shambolic as well. The engineering state tried as hard as it could to hold on to earlier triumphs, until it was forced to let everything go.

After the Shanghai lockdown, it grew increasingly apparent that there was no industry that Xi would hesitate to crush and no personal misery worth noticing by the state if it could halt the spread of omi-cron. It didn't matter that companies were feeling deeply uncertain about future investments, that local governments were running out of funds as they spent everything on testing, and that people were deeply exhausted. When the Italian philosopher Giorgio Agamben wrote in 2020 that his country's pandemic control measures resembled a "sanitation terror" and a "juridical-religious obligation that must be fulfilled at any cost," he was criticized. His remark, in my view, applies in far stronger force to the engineering state's commitment to zero-Covid. Chinese people grew livid that the medical system was prepared to ignore any number of deaths from diabetes, cancer, and other life-threatening conditions and that their entire lives had to be subordinated to the targeting of this number.

Hill peoples in Yunnan and other parts of Zomia have mounted occasional insurrections against various state controls. So I found myself wondering why more people did not attempt to protest Shanghai's lockdown. The United States' patchy lockdowns, mild even by the standards of European countries, produced mass unrest in the summer of 2020. Occasional scuffles had broken out between angry Shanghainese and police, but there was no broad rebellion. Though China's domestic security budget is larger than the budget for its military, the state never even had to uncoil its more fearsome elements like the People's Armed Police to enforce lockdowns. Regular police were all it needed.

A Shanghainese friend helped me appreciate the subtlety of police tactics. He lived in a compound in the French Concession with a lot of foreign residents, becoming, like Owen, one of the unofficial representatives of his building. One day during the lockdown, several of his neighbors acted out their frustrations by toppling a barricade. Afterward, police went through their surveillance videos, identified every perpetrator, and brought them into the station for interrogations that lasted hours. My friend told me that the police rarely asked open-ended questions, saying rather, "Confirm that you kicked the barricade so-and-so many times," and then writing up their statements and demanding their signature. They didn't impose any punishments. But these statements hung over the residents. A freaked-out French couple who signed statements subsequently departed the country for good.

I've asked several friends why they thought Shanghainese did not protest. They wondered that too. The main reason they proposed was that most Chinese were genuinely fearful of catching the virus. They had listened to too many government reports of how virulent it was and few reports from Western commentators that downplayed its seriousness. China's health authorities had adopted a gradualist approach to layering on its measures, such that the zero-Covid strategy did not feel so strange until later. And no one imagined that the lockdown would last as long as it did. People might have protested earlier if they knew that the lockdown would last eight weeks, but the city's initial announcement of an eight-day "pause" forestalled dramatic action.

But Shanghai was tense after the vividness of an eight-week lockdown. Nobody knew how Xi planned to exit from zero-Covid: Wasn't everyone going to catch the virus anyway, potentially meeting it with a domestic vaccine that was less efficacious than what the American government was giving out? Shanghai tightened control over movement restrictions after it reopened in June, announcing that such measures were necessary to prevent another lockdown. For a while, people continued to accept them.

I was just impatient. There would be protests throughout the country in the fall of 2022. In Shanghai, they turned intensely political. I'll never forget that I witnessed open antigovernment demonstrations in China's richest and most populous city.

. . .

SILVIA AND I DEPARTED from Yunnan at the end of the summer. We returned to a Shanghai tense from the fresh trauma of lockdown.

The city's restrictions were more consistently enforced than before. I couldn't enter a public space—the subway, a restaurant, a convenience store—without displaying my health code showing a negative PCR test taken in the past seventy-two hours. The city put up kiosks on many street corners, but it was easy to forget to take a Covid test in time, making it no longer possible to meet one's friend at a restaurant or café. One day, I had a lapse. When I stood on a sidewalk trying to order a coffee from a window counter, I faced the absurdity of being refused. The barista shrugged and turned away when I showed my anger.

In part due to tougher measures, Shanghai did not see rising caseloads through the fall. Omicron, however, was spreading through other cities across the country.

An earthquake struck Sichuan in September 2022. When panicked residents in the city of Chengdu hastened to exit their homes, pandemic control officers barred some people from leaving, locking them inside trembling buildings. A bus carrying people to quarantine facilities overturned on hilly terrain in Guizhou, killing twenty-seven people. A fire broke out in Urumqi, the capital of Xinjiang, where ten people died after fire trucks were obstructed by pandemic-control barricades such that they couldn't direct water on the blaze. All these were extensively reported in Chinese media. So was the 2022 FIFA World Cup, where millions of soccer-loving Chinese watched crowds of people cheer inside Doha's stadiums. Two years ago, they'd scorned how the rest of the world handled the virus. Now Chinese watched with envy and wondered, was Covid really more dangerous than fires and earthquakes?

Xi Jinping wanted nothing to go wrong in 2022. At the party congress that October, he was about to appoint himself to a third term. It would have disrupted his political plans to let Covid break loose in China, triggering the sort of unrest that frightened the leadership at any time, but especially before the party congress that takes place once every five years. Xi grew obsessed with creating a stable political environment.

In May 2022, still in the middle of Shanghai's lockdown, the

Politburo announced that China's zero-Covid policy "can stand the test of history. . . . Just as we have won the great battle for the defense of Wuhan, so too can we triumph in Shanghai." The statement carried a sting for any doubters, promising to "fight against any speech that distorts, questions, or rejects our Covid-control policy." In September, the Internet Monitoring Bureau of the Public Security Ministry issued a directive for propaganda authorities to broadcast only approved messages and to "stop spreading negative energy!"

By the end of October, as China's largest cities escalated their pandemic controls, Xi secured his third term. But after more than two years of tightening controls and of tragedy, people were enraged.

Protests broke out and turned violent at Foxconn factories in Henan. Electronics assembly is exhausting and repetitive at the best of times; for thousands of assembly-line workers making iPhones, the stress became too much. It's not clear what exactly prompted the unrest—missed payments, factory bubbles, or the spread of the virus—but it brought young men onto the streets. Videos showed workers facing off against massed riot police, some of them in their *dabai* suits, throwing bricks, fence segments, and stones into the masses of white uniforms. And the police were retreating.

Protests turned political elsewhere. One man in Chongqing went viral for shouting the American Revolutionary slogan "Give me liberty, or give me death!" Onlookers initially protected him from the police, but authorities eventually managed to shove him in a car. In the bar district of the French Concession, people chanted something far more threatening to the regime. One day in November, people held a vigil on Urumqi Road, the center of Shanghai's bar district where many foreigners live. The road just happened to be named for the city where ten people died in a blaze. It wasn't meant to be a big event. But it gained energy when tipsy young people staggered out of their cocktail bars and joined the commemorations.

At some point after midnight, the initially subdued vigil turned into a protest. Young people began yelling out their frustrations, surrounded by police, though they did nothing to stop their chants: "Down with the Communist Party! Xi Jinping step down!"

It was an impromptu protest, taking place precisely because it was never organized. I had already gone to sleep that night. The next day, I walked twenty minutes west of my home to Urumqi Road. There I found a substantial police presence with a lot of people milling around. I ran, coincidentally, into my friend Owen. The air thrummed with nervousness and excitement. When a car passed by, blasting the Chinese national anthem at high volume, we all perked up to see if the police would do something. We saw one police officer drag away a reporter from the BBC. In the evening, police moved decisively to clear people away from Urumqi Road. We saw them slowly disperse people from the zone, until they put up high barricades throughout the whole street, blocking most sidewalks.

Then there was a more individual protest. One morning, a man disguised as a construction worker draped two banners on a busy highway bridge in Beijing. He then burned a tire to create smoke and draw attention to his words. The first banner, loosely translated, read:

End the tests, we have to eat;
Stop with curfew, we want to be free;
Enough with lies, we demand dignity;
Reject the Cultural Revolution,
Reform and Opening is the solution;
We don't need a great leader, but a free election;
We are citizens, not slaves.

The second banner read, *"Remove the national traitor Xi Jinping."*

Police arrived to arrest him and take down the banners but not before these slogans started spreading on social media. The protester's identity remains unconfirmed. What's certain is that he's paying a grievous price for hanging the banners. Censors have struck the highway bridge from China's mapping services. If you put in "Sitong Bridge," the service says that there is no result.

Some of the youth protesters suffered too. Young people gathered in Shanghai, Beijing, and a few other cities at the end of November, sometimes holding up a blank piece of A4-sized printer paper. Carrying blank pieces of paper became a way to symbolize China's censorship. It was a perfect echo: Whiteness represented the enforcement of pandemic controls, through the protective medical suits of massed groups of *dabai* (big whites), until young people appropriated it for protest. Later, anti-Covid demonstrations in China were collectively known as "the white paper protests."

Youth chanting, hanging slogans, holding blank pieces of paper— these would be paltry acts of protest in any democratic country. But it's hard to overstate how rare it is to see public acts of defiance in China, especially after Xi dedicated immense resources for surveillance and enforcement to smother exactly this sort of demonstration. I certainly would never have expected to hear shouts of "Down with the Communist Party, Xi Jinping step down!" in Shanghai while police haplessly stood by. Even if these are fairly small acts, Beijing's bridge man and the young people in Shanghai deserve to be remembered for their courage.

The number of protesters was never very large. They were special because they involved upper-class Chinese families: wealthy people who didn't want to suffer lockdown and well-off youths who attended good schools. The Communist Party had always counted on these people for their support. The denouement of China's Covid experience features broad exhaustion.

• • •

THROUGHOUT NOVEMBER 2022, while these protests took place, the virus was spinning out of control. Crucially, it wasn't under control in Beijing. As the city moved to lock down, it encountered more resistance among residents. Beijing residents are proud of their heritage of challenging power, and many people were nervous about meeting the same fate that Shanghainese suffered in the spring.

The government's response grew erratic. While top officials in Beijing were insisting that pandemic controls must continue, local cities around the country barely enforced any controls. By the start of December, the state had announced several rounds of "optimization" measures, the last of which abandoned the language of dynamic zero clearing. Nearly three years after it began, zero-Covid was over.

I caught Covid in Shanghai on December 23. It was a mild case, and many other people were less fortunate. The timing was nonsensical: The state had dropped all restrictions during the worst of winter. China hadn't meaningfully accelerated its vaccinations of the population beforehand; it remains mystifying why they didn't give people shots in any of the dozens of times it forced people to take PCR tests. Doctors and nurses received no special warning that zero-Covid would end abruptly, leaving them to face a surge of patients.

When I think back to this moment, it's the lack of fever medications that really sticks with me. For three years, the government made it difficult for people to buy ibuprofen, Advil, and other fever reducers for fear that people might disguise their fevers to avoid detection. During an outbreak, pharmacies limited purchases of fever meds or removed fever meds from their shelves entirely. Therefore, much of the Chinese population met this Covid wave without medication on hand. As best as I can tell, China is the only country that denied its people fever medications during a fever-producing pandemic. It is a perfect encapsulation of the engineering state's twisted logic.

Propaganda authorities had no special warning, though they shifted seamlessly from declaring that the virus must be stomped out in one week to saying that everyone had to be responsible for their own health the next. It felt like living through the scene in Orwell's

1984, in which officials switched directions, mid-speech, declaring that Oceania was at war with Eastasia rather than Eurasia.

Why did Xi suddenly abandon zero-Covid? I don't think that the protests played the biggest role. Far more important was that people throughout the country had grown exhausted by lockdowns, which robbed them of their sanity and livelihoods. Local governments were just as exhausted, with many of them facing financial stress from doing so much testing while forgoing economic activity. (Economists from Nomura estimated that testing cost 1.8 percent of China's GDP in 2022.) When the virus became entrenched in Beijing, I suspect that the central government took a good, hard look at whether it could enforce a lockdown on the capital, which has always enjoyed the greatest political pampering. Local jurisdictions around the country were already abandoning their own controls. Beijing decided that pandemic controls were no longer tenable. And so the virus came.

Xi Jinping didn't make many public appearances between December and January. He never stepped out to explain the reversal of a policy he personally and forcefully insisted on, nor did he attempt to offer much comfort to people who faced an illness that his state spent three years terrifying people about. Crematoriums were operating nonstop from the end of December 2022, though the state failed to announce that many people had died from Covid. It was interesting that the Chinese Academy of Sciences released a burst of obituaries in December 2022 for the senior scholars who had just passed away.

In January, the state mouthpiece Xinhua released a commentary attempting to rebut that the move from zero-Covid to total-Covid was haphazardly planned. Rather, the news agency stated that "all decisions were made after scientific analysis and shrewd calculation" and that these were "by no means impulsive decisions."

The reversal felt too abrupt to be shrewd, but no matter. The twentieth party congress concluded with a full political sweep for Xi. His choice for a new premier (China's head of government) stung many people: Li Qiang, Shanghai's party secretary on whose watch the lockdown took place. Through 2024, every governing party in developed democracies lost vote shares, including the Democratic Party in the United States, as voters tossed out the incumbents they blamed for pandemic management. In authoritarian China, the politician who oversaw the largest lockdown was elevated to the second-highest office.

And so the Covid-19 pandemic ended in China as it began, hostage to political events: willfully ignored by Wuhan authorities in the beginning and then by the central government at the end.

After moving from China to the Yale Law School, I gained some new perspectives. It was a good thing that the United States stumbled to "live with the virus." I found one item particularly quite irksome on my return to America in 2023: a yard sign that begins "In this home we believe science is real." The Communist Party "followed the science" of zero-Covid to its logical conclusion: barring people from their homes, testing people on a near-daily basis, and doing everything else it could to break the chains of transmission. Four decades ago, it "followed the science" to forcibly prevent many pregnancies in the pursuit of the one-child policy.

We can agree that "science is real." But we have to keep in mind that there is a political determination involved with how to interpret the science. And that is something the lawyerly society is better at. It has lawyers interested in protecting rights, economists able to think through social science, humanists who consider ethics, and many other voices in the mix, attempting to open policy prescriptions up for debate. China doesn't have a robust system for political contestation; engineers will simply follow the science until it leads to social immiseration.

Engineering only works if it is using good data. But data probity is another of China's casualties in the aftermath of Covid. The government has had a wobbly commitment to accurate reporting at the best of times. After the pandemic, the government has more regularly succumbed to the temptation not to share bad news. China announced a total of around 125,000 deaths related to Covid-19, an absurd undercount when scholarly estimates come to nearly 2 million excess deaths. After 2023, China is fudging many other pieces of data, from birth rates to youth unemployment.

Imagine the comedy romp that could be produced about colleagues who used to despise each other learning to come to terms while they were stuck for two weeks, unable to wash, inside their offices. Or the relationship drama between a couple at Disneyland, attempting to resolve their problems as they remained unable to depart from the happiest place on earth. Unfortunately, the state has suppressed any official memory of Shanghai's lockdown itself. The engineers want people to forget, not to poke fun at this experience.

After zero-Covid, Shanghai is a little bit less like the Paris of the East, a little bit more like another Pyongyang. The city remains amazingly beautiful, with so much art deco, neoclassical, and modernist architecture. Its pleasures continue to deepen, with entrepreneurs competing ferociously to introduce new ways to have fun. But it also has long-term wounds that are not so visible. Owen, who is still in Shanghai, told me that lockdowns no longer come up often in conversation. "But when people get really drunk, it's still something that people get worked up about."

My friends felt like they were taken twice to the cleaners: first, when they couldn't stockpile essential supplies following the surprise lockdown announcement and, later, when they couldn't stockpile any medicine. What was the point of the April–May lockdown, they ask, when it was all given up just nine months later? Some of the older people said that the lockdown wasn't the worst thing to happen to them, pointing to the Cultural Revolution. Younger people born after 1990, however, who had known only rising prosperity, had their first real taste of the disaster that could be inflicted by the engineering state.

The Shanghainese elites I knew had a crisis of faith. None of them quite imagined that the spikiest, most coercive instruments of the Chinese state could be pointed directly at them. Nationalist tongues stilled to silence, for a while, after they wagged for two years about how China's pandemic controls proved its superiority over the West. No wonder business dynamism has fallen in China's richest and most cosmopolitan city.

The three years of pandemic controls allowed Xi Jinping to indulge in central planning, not only to express certain egalitarian ideals embedded in common prosperity but also to control physical movement of millions. The one-child policy brought the Communist Party to reach deep into women's bodies; the digital surveillance developed as part of zero-Covid has allowed it to control even a person's daily access to her shower. There's now a direct institutional linkage between the two policies. The neighborhood committees that took a starring role in enforcing Covid lockdowns haven't been disbanded; they are now being used to call up recently married women to ask about their menstrual cycles and whether they wouldn't like to have a few children. Some are able to bear it. But many young Chinese are tired of being lectured by old men to work hard and have kids while facing a horrid job market.

CHAPTER 6

FORTRESS CHINA

THE MOST REMARKABLE NEW Chinese slang word that developed during the pandemic was *rùn*.

It means what it sounds like. Chinese have appropriated this word (meaning "to moisten") for its English meaning to express their desire to flee. Throughout the unpredictable and protracted lockdowns, *rùn* evolved to mean leaving big cities, where pandemic controls were tightest. Or it meant emigrating from China altogether. After I departed from China in 2023, I kept meeting Chinese who have, in recent years, decided to emigrate, gambling that their lives would be better abroad.

Young people want to go to Europe, the United States, or an anglophone country, but these governments tend to be miserly with visas to Chinese. Thus, many émigrés go to nearby countries in Asia. Those with ambition and entrepreneurial energy flock to Singapore, where Chinese companies like ByteDance have set up big offices. Those with wealth and means buy themselves a pleasant life in Japan.

Everyone else—slackers, free spirits, kids who want to chill—is hanging out in Thailand.

At the end of 2023, I spent a month in Thailand's Chiang Mai with people who have *rùn*. I had gotten to know many of them the prior year while I lived in Yunnan. They were young, creative types. These people working in journalism, the arts, or tech went to this mountainous part of southwestern China after they felt stifled by lockdowns and political controls on speech. Yunnan officials tended to be more relaxed, looking the other way while these youths immersed themselves in cryptocurrency projects by day and relaxed at speakeasies at night. These free spirits were interesting to me as a counterpoint to the cultural mainstream. They reject the corporate grind of Beijing and Shenzhen. They want to invent their own lives.

But even Yunnan has grown more restrictive in recent years. Some of these people therefore took a plane to hop over the mountain ranges separating the province from Thailand. Why Thailand? Because it's easy. Chinese can visit for short stays without a visa; a longer-term residence isn't difficult to arrange. If they so much as sign up for language classes or Muay Thai boxing lessons, they could qualify for an educational visa. None of them take these educational demands, or life itself, all that seriously. They are in their twenties or early thirties, trying to catch up on three years of lost fun after zero-Covid.

Many of them practiced the intense spirituality possible in Thailand. Chiang Mai is a beautiful holy city studded with golden-roofed temples and monasteries. One can find a meditation retreat in these temples or go to even more secluded retreats in nearby mountains. At these places, one meditates in silence for up to fourteen hours a day, speaking only to the head monk each morning to tell him about the previous day's breathing exercises and hear the next set of instructions. After twenty straight days of this regimen, one person told me that he found himself slipping in and out of hallucinogenic experiences.

The other spiritual wellspring comes from the use of actual

psychedelics, which are easy to find in Chiang Mai. Thailand was the first country in Asia to decriminalize marijuana, where pot stores are nearly as common as coffee shops. It seemed like everyone had a story about using mushrooms, ayahuasca, or even stronger magic. The best psychedelic mushrooms are supposed to grow in elephant dung, leading to a story I heard of a legendary set of backpackers who have been hopping from one dung heap to another on a long, unbroken trip.

I spoke with the young Chinese who are in Chiang Mai to have a good time and also with the longer-term residents about why they've decided to reside there. None of them made the decision to emigrate easily.

Yiju was one of the people starting over in Chiang Mai. He's a friendly software developer in his thirties, pudgy from too much time in front of a computer screen. He worked for a while in Silicon Valley before he found himself caught up in the cryptocurrency craze in 2018. So he returned to China at a time when it was a major hub of cryptocurrency activities. Like many people in crypto, Yiju embodies a certain idealism. That came through with his eagerness to announce his views in manifestos filled with punchy statements on how the economy should operate and how people need to be kinder. Unlike many people in crypto, he was also given to quiet reflection on the limits of technology as well as what China means to him.

"China feels like a space in which the ceiling keeps getting lower," Yiju told me one day. "To stay means that we have to walk around with our heads lowered and our backs hunched."

Young people in Chiang Mai told me they've felt a quiet shattering of their worldviews over the decade of Xi Jinping's rule. These are people who grew up in bigger cities and attended good universities, some of them overseas, which endowed them with certain expectations: that they could pursue meaningful careers, that society would gain greater freedoms, and that China would continue to be more integrated with the rest of the world. These aspirations have mostly shriveled. Though their lives in big cities can be quite pleasant, with new milk tea shops to try or art spaces to take selfies around, they work in jobs that are stressful and menial. They feel smothered by political controls. After the lockdowns, many of them grew aware that they had an unwelcome tendency to inflect every future scenario with a sense of catastrophe.

Not everyone has been thrilled with the move to Thailand, where they don't foresee great job prospects. They haven't all mustered the courage to tell their parents where they really are: Mom and Dad are under the impression that they're studying abroad in Europe. That can lead to elaborate games in order to maintain the subterfuge, like drawing curtains to darken the room when they video chat with family, since they're supposed to be in a totally different time zone, or keeping up with weather conditions in the city where they're supposed to be so that they're not surprised when parents ask about rain or snow.

Yiju fled in the wake of the white paper protests against Covid. When police sought him out for questioning, he went to hide in a monastery. Many of the other residents in Chiang Mai had participated in the protests against Covid restrictions and have had friends who were arrested. Everyone had experienced some alienation. A few lost their jobs in Beijing's crackdown on digital platforms. Several had worked in domestic Chinese media, seriously disgruntled by censors. Writers in particular have a hard time dealing with the shock of working for months on a story only for censors to delete it hours after publication. The first time that happens you're enraged, the second time you're embittered, the third time you *run*.

In Chiang Mai, these creative types gathered around a bookstore founded by a journalist. Nowhere Books had its first store in Taiwan before opening a second branch in Chiang Mai, offering books that can't be bought on the mainland. Nowhere's references to politics are subtle. Mixed with popular books—novels, travel guides, cookbooks—are works by authors that couldn't possibly be published in mainland

China. The store is proud to carry a Chinese translation of the *Whole Earth Catalog*, the Californian counterculture magazine published through the late 1960s and early '70s that advocated for each reader to "conduct his own education." And around the bookstore are faintly subversive signs: a sticker of the Urumqi Road sign, the focal point of Shanghai's protests, and jesting passports handed out by the bookstore inviting patrons to become citizens of the Republic of Nowhere.

Many of my friends, both Chinese and foreign, have *rùn* too.

Shanghai's foreign population was in decline even before the pandemic: Between 2010 and 2020, China's most internationalized city lost a quarter of its long-term foreign residents. Since the lockdowns, this population has taken another big drop. Shanghai drew foreigners and Chinese who were excited about the economic and creative boom in the city. For business executives, a posting to China used to pave the way toward the C-suite. That's starting to feel less the case since China has become such a different market (given political complexities and data controls) that a posting there is now viewed as a quagmire. As China's economy slowed, people wondered why they were living in a place with uncertain growth and a lot of drama.

Xi might not be so upset with the creative types who want out. He might not be much bothered either by the foreign expats leaving Shanghai, even if they work at important companies like Apple or Tesla. But Beijing has displayed greater concern about the number of rich people taking their money out of the country.

My friend Jessie is the daughter in a wealthy family, growing up in both her native Shanghai and Vancouver. A tall girl with curly hair, she has usually been more interested in frequenting fitness classes than in reading the news. She had previously not paid much attention to political events, feeling like it wasn't worth her while to dwell on matters that were often gloomy and always impenetrable.

Then she lived through Shanghai's two-month lockdown. Subsequently, Jessie started to follow politics. "This stuff could affect us, you know," Jessie told me one day while she was visiting in New York. She was talking about the Third Plenum meeting of the Central Committee.

"Could it?" I asked, surprised to hear that she was monitoring this weeklong party gathering.

"You never know what they're going to do," Jessie said. When I asked her whether party announcements have ever prompted her to act, she replied no. Paying attention is already a novel activity for her. It might, I feel, lead to more active political engagement in the future.

Jessie is keeping her roots in Shanghai, although she plans to spend an increasing amount of her time in Vancouver. Many other wealthy Chinese have decided to settle elsewhere for good. Hard numbers are difficult to work out, but one UK-based emigration firm estimated that nearly 14,000 millionaires emigrated from China in 2023 and over 15,000 in 2024. Parts of the United States popular with Chinese, like Irvine, California, have seen a surge in new homebuyers. Both the United States and Canada have reported a doubling in the number of Chinese migrants who have obtained permanent residence after making a large investment (which could mean buying property): from 2,000 to 4,000 in Canada between 2019 and 2023, and from 3,900 to 7,500 in the United States between 2019 and 2024.

Less fortunate Chinese take a different path to the United States: through a grueling trek across the southwest border. US border officials have apprehended rising numbers of Chinese nationals: from 450 in 2021 rocketing to 38,000 in 2024. The flow diminished in the second half of 2024 due to tighter border enforcement. But still there have been more than a thousand Chinese nationals attempting to cross the border on foot each month for two years. Many have flown to Ecuador (which did not demand a visa from Chinese nationals until July 2024) and then have taken the perilous road through the Darién Gap.

The creative diaspora has launched cultural events in the United States. New York and Washington, DC, have new Chinese bookstores like Nowhere in Chiang Mai. Once a month in New York, a feminist group operates an open-mic show for comedians to perform their acts in Mandarin. They sell out so quickly that I was lucky to get a ticket. On a chilly day in October, I went to an Italian restaurant in Midtown Manhattan that rented its basement for shows. Around a hundred people gathered that day to hear ten women performing a "story slam" rather than the usual standup. One person spoke about how she connived her way into an exclusive Berlin nightclub, and several shared accounts of their dating lives. Most stories tended to be sad: dealing with a layoff or the death of a grandmother. The audience reacted with tremendous encouragement whenever the performers' voices cracked or their storytelling faltered.

A decade ago, it might have been difficult to imagine that New York would have a set of feminists organizing standup in Mandarin through which runs a streak of political discontent. As Xi became a more assertive leader, more Chinese have become unhappy with China's direction. What is most surprising is that desperate migrants are willing to abandon the "China Dream" that Xi has preached and that they are willing to embark on a dangerous, monthslong journey to cross the US southwestern border.

Why are so many Chinese still leaving? Because entire generations feel whipsawed by the engineering state's violent mood swings. Their jobs, and indeed their lives, in China felt like dead ends. They're not making great money in Thailand either, but they are able to have a lot of fun in its relaxed atmosphere.

Xi has talked about achieving national greatness without backing it up with economic growth. The trouble is that when people suffer—as they do through a property collapse, high unemployment, or lockdowns—they start to wonder what they are really getting. It's certainly not enrichment. When they're given a cold, hard smack in the face by something that certainly doesn't feel like greatness, they become unmoored. This sense of alienation has been a big reason to *rùn*.

• • •

AFTER SIX YEARS in China, I missed pluralism. It is wonderful to be in a society made up of many voices, not only an official register meant to speak over all the rest. I missed the ambient friendliness of Americans combined with a government that mostly leaves people alone. Most of all, I missed the ability to order books. To be able to read physical books, I relied on my folks to mail me periodic packages, usually in batches of twenty kilograms, while accepting the uncertainty that any of them might be confiscated by customs agents. It heightened the physical ecstasy of opening the box to see how many passed through the censor's gauntlet. But it was a thrill I could have lived without.

So I had *rùn* myself after the collapse of zero-Covid, when I moved from Shanghai to Yale Law School. Shanghai has many things superior to that of any American city: walkable and safe streets, vibrant street life, splendid food, an ease to go anywhere in the city or the country through mass transit. It was the Chinese government's overbearing presence—censorship, intolerance of dissent, a lingering threat of catastrophe—that pushed me away. The operators of the Great Firewall decided that my little personal website, where I publish my annual letters, should be blocked. I am still puzzled.

I changed my mind about several things over my time in China.

When I moved to Hong Kong at the start of 2017, I entertained the idea that we were living at the start of an "Asian Century," in which China and India would restore Asia to the economically dominant role it played centuries ago. I didn't believe it, necessarily. But it didn't feel like a crazy scenario. Donald Trump, after all, had been shooting admiring glances at autocratic countries while unloading his petulance on Canada, Europe, as well as other American allies. Xi, by contrast, displayed a patient resolve to strengthen Chinese capabilities. Parts of that remain real, although I now have a better appreciation of China's weaknesses. There are many things that China will

be successful at, but I departed the country with a better appreciation of the self-limiting features of the Chinese system. Most notably,
the Communist Party distrusts and fears the Chinese people, limiting
their potential for flourishing.

The engineering state tends to begin impressively and end disas-
trously. The pursuit of zero-Covid isn't the only example of that tendency I lived through. The regulatory storm that Xi unleashed against
China's digital platforms is another case in point.

In May 2024, while attending a symposium of entrepreneurs and
investors in Shandong province, Xi Jinping asked the group, "Why are
we producing fewer and fewer unicorns?" This stray comment created
a minor ripple online. Why is China no longer a leader in producing
the sorts of tech start-ups that are valued over $1 billion? Before their
comments were censored, people posted, "But sir, you are the cause";
"Is the leadership compound in Beijing connected to the Internet?";
and "They were frightened away by blank pieces of paper."

Xi's question had produced fresh worry among businesses.
Authoritarian systems aren't good at disseminating bad news. The
coronavirus had spread, after all, because local officials in Wuhan
refused to let the news of a virus disturb their political serenity as they
arrested medical whistleblowers. Companies and investors therefore
wondered whether Xi was genuinely unaware of how much his policies had destroyed major segments of the economy. Perhaps nobody
had told Xi that he was the most feared unicorn hunter of all.

For a while, China produced a herd so lusty that it looked like
they were on the verge of outpacing even the unicorns in Silicon Valley. They raced neck and neck against their American counterparts
in e-commerce, ride hailing, and social media. Sometimes they had
help from Beijing—most notably when the state drove out Google
and Facebook to the benefit of local platforms like Baidu and Tencent. Sometimes they outcompeted American firms, like Amazon and
Uber, more or less fairly through brutal wars of maneuver. ByteDance

had created a new category of short-video apps with TikTok, while new e-commerce platforms sprang up to challenge Alibaba. That company's flamboyant founder, Jack Ma, would have fit in among the more eccentric personalities from Silicon Valley.

During this era of light regulation, China's unicorns grew into mighty beasts. Lu Wei was the director of the Cyberspace Administration, making him the chief internet regulator. He was a colorful character in that role. When I visited start-ups in Beijing around 2018, I heard lurid stories: Lu supposedly took equity in companies and then placed his regulatory thumb on the scales in their favor; sometimes, he would walk through an office and remark on how pretty a female employee was, expecting executives to take his hint. His reign was characterized by regulatory forbearance, perhaps because he was a personal beneficiary of the sector's growth.

In 2018, Lu fell from grace. The Central Commission for Discipline Inspection expelled Lu from the Communist Party and published an unusually explicit list of his crimes. It went beyond the usual accusation of bribery to include charges of "deceiving the central leadership" and "trading power for sex." Lu subsequently wrote a letter so self-abasing that it was featured in a national museum celebrating forty years of China's policy of reform and opening.

China's tech companies were on the verge of convincing global investors that they could reach the valuations of Silicon Valley giants. At home, however, they produced similar forms of discontent as their American counterparts, facing allegations of exercising corporate power against smaller firms and insufficiently protecting data. Lu Wei's fall took the era of light regulation down with him.

New regulators subsequently announced that digital platforms would be subject to "rectification measures." A former ByteDance executive has publicly accused the company of facilitating bribes to Lu. And ByteDance became the target of an investigation, which would later produce a groveling public apology from that company's founder. "I have been filled with remorse and guilt, entirely unable to sleep," Zhang Yiming, then CEO, wrote to his staff. "Our product has been incommensurate with socialist core values. . . . I am responsible because I failed to live up to the guidance and expectations supervisory organs have demanded."

But China's tech platforms continued to grow larger, developing certain digital capabilities that the state did not have and barely understood. Ominous rumblings emerged from the central leadership. Xi issued warnings against the "disorderly expansion of capital" and promised to "deepen structural reforms." Starting in late 2020, Beijing declared open season on the digital economy. Every government agency lined up to take shots.

Securities regulators derailed the public listing of Ant Financial, a fintech company founded by Jack Ma, accusing it of sowing financial instability. Data regulators investigated Didi, a ridesharing app that had just gone public on the New York Stock Exchange, for vague charges of endangering national security. The press regulator announced that minors were permitted to play video games during only three designated hours per week: between 8:00 and 9:00 p.m. on Friday, Saturday, and Sunday. Antitrust authorities launched a flurry of investigations against big platforms. Even the Ministry of Education took part in the great hunt: It declared that the online education sector, which offered supplementary lessons outside the formal schooling system, could no longer produce profit.

Over the course of 2021, hardly any major Chinese tech company emerged unscathed. Xi's regulatory storm wiped out a trillion dollars of market value from Chinese companies. New Oriental, one of the education companies, lost 90 percent of its market cap and then laid off 60 percent of its employees. Alibaba toppled from being an $800 billion company to just a quarter of that size two years later. Jack Ma disappeared from public view for months after the cancellation of Ant Financial's IPO. Meanwhile, securities regulators in both the

United States and China were making it more difficult for companies to be publicly listed. And Xi's pursuit of zero-Covid pulverized service industries targeted by tech companies. The economy that emerged out of the pandemic is characterized by high youth unemployment, shaky household confidence, and limp consumer demand.

Unicorns aren't easily bred on such impoverished fields. Especially not when there's a giant hunter stalking to ensure they conform to socialist core values. Consequently, fewer entrepreneurs are founding start-ups, and venture investment in China has collapsed.

Xi's reining in of tech giants are not altogether different from what a lot of American and European regulators wish to do to Silicon Valley. Every government in the world is grappling with companies that have too much influence over the flow of information and commerce. Individually, China's regulations around antitrust, data protection, or financial risks may pass muster on technocratic grounds. But Beijing issued regulations with a speed and ferocity that no other state can match. It did so for reasons that the West would not: to shift investment and talent into state-prioritized industries and to crush the power that these companies were gaining at the expense of the state.

That's another way that the US and Chinese political systems are inversions of each other. In the United States, the political drama is around legislative processes and Supreme Court rulings; implementation of policy is quickly forgotten as political attention moves to the next big issue. In China, the policymaking process is conducted significantly in secret, then its outcome is dumped on the people.

Whereas the United States or Europe might tussle with a Silicon Valley tech giant for years in court and then extract a few billion dollars in fines, Chinese companies don't challenge administrative actions. Rather, they issue supine statements as the founder of ByteDance did, or as Didi wrote after it received an enormous fine, "We sincerely thank the relevant authorities for their inspection and guidance."

The regulations weren't only an exercise of technocratic governance. They added up to a sweeping exertion of political control. China's crackdown consisted of both technocratic regulation *and* an effort to impose political discipline on a freewheeling sector. Xi has forcefully reminded China's tech companies that they cannot represent a power center that challenges the state's sovereignty. It was, in other words, an attempt to change the cultural mindset of companies. The Communist Party reminded them that it retains the discretionary power to engineer all aspects of society, which means putting tech companies in their place.

There might be something to be said for this sort of approach. What if, say, the US government had responded to the 2008 financial crisis by reshaping Wall Street's risk management culture rather than engaging in the endless negotiations that yielded a 2,300-page statute that nobody understands? But Xi's attempt to achieve cultural change has left people disgruntled and whole industries disfigured.

The trouble with Xi Jinping is that he is perhaps 60 percent correct on everything.* He's driving toward a usually admirable long-term goal. But in the name of achieving change, the engineering state delivers such beatings on people or industries that they are unable to pick themselves back up again. Even if Xi's judgments are right, his brute-force solutions reliably worsen things. Does big tech have too much power? Fine, but stomping out their businesses has traumatized entrepreneurs. Are housing developers taking on too much debt? Yes, but driving many of them toward default subsequently triggered a collapse in homebuyer confidence, prolonging a property slump. Does the government need to rein in corruption? Definitely, but Xi has terrorized the bureaucracy to the point of paralysis.

---

* I chose this number deliberately. Deng Xiaoping came up with a formulation that Mao Zedong was 70 percent correct and 30 percent wrong. I am sure that Xi would be the last person in the world who would say that he's greater than Mao. Therefore I assign him a slightly lower score here.

Sometimes, the only thing scarier than China's problems are Beijing's solutions.

That is one of the defining characteristics of the engineering state. The Chinese government often resembles a crew of skilled firefighters who douse blazes they themselves ignited. China's national effort contained the spread of Covid, for a while, after Wuhan officials did nothing to prevent it. Decades earlier, the engineering state overreacted to its population growth with the one-child policy. Economic confidence wouldn't be so fragile if it weren't for regulatory thunderclaps emerging from Beijing.

Here is where the lawyerly society shines. We don't have to worry about the US government imposing the one-child policy or zero-Covid, because it would never with the former and could never with the latter. The United States also wouldn't have caged so many of its tech companies. Lawyers, as I wrote in my introduction, are excellent servants of the rich. Chinese tech founders (and their investors) are indeed very rich. Given the absence of lawyers and a political culture sympathetic to rights, they could find no protection.

After alienating so many people, has Xi decided to change course? No, he's doubling down on promoting engineers to leadership. When Xi coronated himself as China's leader for a third term in 2022, he unveiled a new leadership team stacked with executives of China's aerospace and defense industries. They are people with practical experience managing megaprojects. Yuan Jiajun, chief designer of China's crewed space program, became party secretary of Chongqing; Li Ganjie, a nuclear engineer, became the party's chief personnel manager; and Zhang Guoqing, a former executive of one of China's largest defense contractors, became a vice premier.

Social engineering will increase as well. In 2018, Xi praised teachers as engineers of the soul, a phrase first used by Joseph Stalin a century ago. Xi's instructions have increasingly moved toward physicality. He has talked about how love of the party and the country needs to start young, which means to "grab little ones from the cradle." The party's messages need to "enter the mind, enter the heart, and enter the hands." Beijing's public security office has promised to get up close and personal in its attempts to offer "zero-distance service." These efforts don't sound less sinister in Chinese than they do in any language.

· · ·

SINCE XI STARTED HIS third term in 2022, he has warned ever more darkly about "extreme" scenarios. In speeches to China's national security community, he has spoken about "ensuring normal operation of the national economy under extreme circumstances." What does that mean? As usual, the top leader is oblique, but it suggests that he's worried that China will one day be cut off from the rest of the world. "We must be prepared for worst-case and extreme scenarios," Xi said in 2023. "And be ready to withstand the major test of high winds, choppy waters, and even dangerous storms." So he has surrounded himself with executives from the aerospace and defense agencies. The intention, it feels to me, is to build China into a great fortress.

What sort of dangerous storm is Xi preparing for? Probably outright conflict with the West. Under Xi's leadership, the engineering state is working seriously to harden itself to win a war, should one ever come.

Xi has already put up higher walls. In 2018, while I was living in Hong Kong, I started to tell people that China might close its doors in forty years, by the centenary of the founding of the People's Republic. At that point, it will once again become the Celestial Empire, its people serenely untroubled by the turmoils of barbarians beyond its borders. Most of my friends reacted with disbelief, saying that it was unimaginable to close a country once it has globalized. It turned out that I was off by a centenary: China had been mostly shut in 2021, a hundred years after the founding of the Communist Party. The pandemic was like a practice run—an exercise in what life in China would be like with its doors closed to the outside world. Xi apparently liked what he saw. After the pandemic, Xi has doubled down on self-reliance.

One of the things I've been surprised by in recent years is how many Americans who used to make regular trips no longer care to visit China. These were businesspeople, investors, and academics who are familiar with China. Many of them felt real fear that they would no longer be able to depart once they entered. Most have nothing to worry about, I'm sure. But it is hard to put that fear away after China took two Canadians hostage and after it has imposed so many exit bans on foreign nationals over business disputes or drug charges. Even those who are not afraid of detention cite the annoyance of having their digital lives cut off. Without a VPN, an American traveling to China will have a hard time communicating with her family back home (since many messaging and email apps are blocked), a hard time glancing at news headlines from the *New York Times* or *Wall Street Journal*, and a hard time navigating cities without Chinese payment apps.

I returned to China only once after the dissolution of zero-Covid. At the end of 2024, the country felt more fortresslike than before the pandemic. Shanghai is strangely muted, restaurants substantially less full, the shopping districts lacking vitality. Consumers clearly have less spending power. People have felt profound economic uncertainty after the economy failed to pick back up following the end of Covid controls in 2022. It's not encouraging for the future of Chinese and American relations that there are only about a thousand American students studying in China. Just before the pandemic, there were ten times that many.

China had been moving away from the West. When the Communist Party selected Xi to be general secretary in 2012, the party had reached an important decision: that China would not attempt to try to be like the United States. The financial crisis that started on Wall Street in the preceding years had disturbed China's leaders. Should

China really adopt a system prone to such instability? Around this time, they settled a debate about constitutionalism. Previously, some Chinese legal scholars attempted to advance the notion that the Communist Party should be bound by laws. Lawyers had won striking victories in the protection of individual liberties, garnering significant domestic media as they did so. Then the victories slowed. And the chief justice of China's supreme court publicly denounced the idea of judicial independence, an action that elevated the party above the law. It is clear, in retrospect, that the selection of Xi was part of a course set by the Communist Party not to follow in the United States' footsteps.

China's economy is faltering while the central government becomes more repressive. It is facing more problems around debt, hostile diplomatic relations with the West, and demographic decline, which was a problem even before many attempted to emigrate. All of this is exacerbated by an unpredictable political factor: Aging autocrats easily get cranky, which is a problem since Xi is likely to stay in office into his eighties.

Is the Asian Century still on? Questions involving Asia's future are more subtle and more interesting than who "wins." Even though I don't believe that China will meaningfully surpass the United States as a global power, it still represents a terrific challenge.

The engineering state remains incredibly capable. Though Xi Jinping has grown increasingly comfortable with disregarding economic growth in favor of national security, it doesn't mean the country has turned into North Korea. Chinese firms are still operating in a robust business environment, though one that is definitely more constrained. The country's relations with the West are not so friendly, but there will still be trade and educational exchange. And China will still be a giant market with enormous numbers of ambitious people who want to make their mark. Only now, it is steadily working to insulate itself from a turbulent world filled with conflict.

The engineering state still has many strengths. There is one thing I haven't changed my mind about since 2017: I remain more confident than ever that China will become a technological leader in manufacturing industries.

Marxists like to reason through contradictions. What is the central contradiction facing China? I submit that we must reconcile two realities when we read the headlines. First, the rich, the creative, and the desperate have chosen to *rùn* from the economic and political gloom that pervades Xi's third term. Second, the manufacturing sector continues to go from strength to strength in the mastery of electric vehicles, clean technology, and other advanced technologies.

How might they be reconciled? With the idea of the engineering state.

The reckless interventions that engineers have dealt to economy and society have left many people seriously disgruntled, spurring them to move their wealth or themselves abroad. Meanwhile, China has embarked on a quest to build a technologically powerful country. On that, I believe it might succeed. My view is that Xi will not achieve his bigger gambit, which is to propel China to displace the United States as the world's preeminent nation, measured not only by economic size but also diplomatic influence, cultural output, and national prestige. The control neurosis of the engineers is the fundamental limit to China's power. But it will also push China to be an advanced manufacturer with dominant positions in many of the high-tech supply chains of the twenty-first century, with military capacity to match and a good chance to challenge US hegemony in Asia.

Engineers are bad at several things. They're not very good, for example, at producing appealing cultural products.

During the height of the pandemic, Xi declared that China needs to become more "lovable." The country's image had suffered as people around the world blamed China for the spread of the virus. But China faced a more fundamental problem than the pandemic. Over the past forty years, the engineering state has done a terrible job of creating cultural output that the rest of the world finds appealing.

I regularly ask Americans what sort of Chinese cultural products they enjoy. Even cosmopolitan people need to take a moment to ponder. Go on, think about it. The answers tend to be niche. People cite the movies of Zhang Yimou, who directed *Raise the Red Lantern*, while the more art house–inclined bring up Jia Zhangke. Those who read sci-fi are likely to mention Liu Cixin's *Three-Body Problem*. TikTok might be another reply, although I'm not sure how much that counts, since the app doesn't often serve Chinese content overseas. Collectors of modern art and video gamers tend to have more to say. Broadly, however, most Americans don't seek out music, art, movies, or literature from China.

It's not, I think, out of prejudice. Americans have had no problem embracing the cultural products of East Asia. Japan produced a wave of pop culture that included animé and manga, Nobel Prize–winning novelists, and popular consumer products like the Sony Walkman and Nintendo Gameboy. South Korea continues to churn out hits, whether these are pop bands or breakouts like *Parasite* or *Squid Game*. Chinese youths are as likely to watch a Korean drama or a Hollywood movie as they are to pick a domestic equivalent.

Four decades after China liberalized, its contributions to global cultures are mostly confined to artistic fringes. It is because engineers don't know how to persuade. The Communist Party insists on a history in which the party is always correct and where all errors come from traitors or foreigners. Rather than acknowledge fault and tell persuasive stories, the instinct of the engineering state is simply to censor alternative narratives. Xi comes across as someone who is a little bit too eager for groveling respect from the rest of the world, which is exactly why he'll never get it.

The issue isn't that Chinese people are somehow less imaginative. Rather, the state's deadening hand has suppressed their creativity. I know that Chinese kids are creative and capable of driving a surge of lovable culture if only they didn't have to face an overbearing censor. After a stand-up comic in Beijing made a joke in 2023 that deployed a military slogan as the punchline, censors crushed the comedy industry. The comedian, Li Haoshi, was detained, his social media platforms suspended, and the studio that employed him fined $2 million. Comedy troupes, which have to submit their scripts to censors weeks before any performance, found their shows canceled throughout the country. For months afterward, comedy clubs across Shanghai were closed.

Engineers can't take a joke. It's hard for art to thrive in an atmosphere of political paranoia plus social control. Today, Chinese artists and writers have to follow socialist core values, which cannot carry a whiff of political criticism. Directors are finding their movies inexplicably pulled from theaters or international film festivals. Most of the movies released domestically are nationalist blockbusters, sappy romances, or supernatural action flicks. No wonder these aren't exportable. Even among captive Chinese audiences, they're not necessarily popular.

The Communist Party's Propaganda Department has treated the media like a manicured garden. It has walled out a lot of foreign content, blocking access to Wikipedia, social media, and many news sites. Only a handful of Hollywood films a year are approved for showing in domestic theaters. Artists know they have to trim their content to be in line with political sensibilities or be uprooted. And propaganda authorities spend a lot of effort to bolster official voices. Step into a bookstore in China and very likely the desk at the entrance will feature a table full of collections of Xi's essays, pristinely arranged (and mostly untouched). Go into a museum and you might find one of his quotes plastered on the wall, having nothing to do with any exhibition. Even aggregators that are not run by the state, including ByteDance, always reserve the prominent spaces for messages directed by propaganda authorities.

The control neurosis of engineers is also an obstacle to another characteristic of a great power: a global currency. The US dollar is overwhelmingly the world's dominant currency, while China's renminbi accounts for 3 percent of global payments. That share has barely grown over a decade. Beijing has imposed a stiff system of capital controls to prevent money from easily moving out, which promises greater stability for the country's highly leveraged financial system. These are exactly the sorts of restrictions that are anathema to global financial institutions. So long as Beijing insists on capital controls, there's a ceiling on how much the rest of the world will want its currency.

China's rise has faltered for many reasons. But there is one thing that it has continued to do well. What do engineers like to do? Build. That has produced considerable benefits at home for spreading material benefits throughout the country, even in very poor provinces. It has helped build food and energy resilience throughout the economy. The engineering state is still on track to become an advanced manufacturer that dominates most of the tech supply chains of the twenty-first century. And the focus on building is winning China some degree of support in developing countries as well.

Exporting China's infrastructure is core to the Belt and Road Initiative (BRI), one of Xi's signature initiatives. Chinese firms have taken their expertise in building roads, bridges, railways, tunnels, dams, and power plants abroad. And they sometimes also bring the sorts of surveillance systems and censorship tools that find eager customers among autocratic leaders. They have gone on a spending spree overseas, with $1 trillion worth of loans outstanding in 150 countries. China has financed trains in Southeast Asia, ports in Europe, light rail in Africa, roads, bridges, libraries, sports stadiums, and many other things besides. According to Deloitte, China has become the single largest financier of infrastructure in Africa, building one in four projects on the continent.

Its results are mixed. Some of the infrastructure projects have helped cement China as a trade hub: Its high-speed rail link with neighboring Laos, for example, has facilitated exports and investment. But not even Chinese construction firms are immune to cost overruns and project delays when they build abroad. One of the BRI's flagship projects is a high-speed rail line connecting Indonesia's capital Jakarta with the city of Bandung. Though the railway has high use, Chinese builders went a billion dollars over budget and completed it four years late. Locals have complained that BRI projects tend to bring the entire workforce from China. Several countries that signed on to the initiative have since withdrawn, most notably Italy. Two photographs have circulated on the Chinese internet: of the Belt and Road Forum in 2017, when Xi Jinping was surrounded by 120 world leaders, and of the same forum in 2023, when there were only three dozen.

Even if Chinese construction companies haven't always shown consistent respect for foreign workers and the local environment, and even though several Belt and Road countries are clamoring for debt forgiveness from Beijing, it appears to have been a net positive for China. On a narrow financial view, the World Bank found in 2024 that BRI projects have generated a positive return for Chinese lenders, though it is small. China has built useful infrastructure in countries that need it. So it's not surprising that overall, developing countries hold China in more positive regard than do Americans and Europeans.

So China's strategy has been to try to rally the rest of the developing world to its side. Proponents would say that China might not need great relations with the West when there are billions more people in the developing world, who have higher economic growth rates than the United States and Europe, which is all true. But consumers in Africa, Southeast Asia, and Latin America have far less spending power than Europeans. And Chinese firms will have a harder time becoming global leaders when they're barred from selling to richer consumers, giving them the profits to compete with incumbents that can. Meanwhile, diplomatic relationships are rarely uncomplicated, as manufacturers in developing countries have suffered from Chinese exports too. Officials from Brazil, India, Indonesia, and South Africa have all pleaded with Beijing to have a more balanced trade relationship.

There's one more thing that engineers are especially good at: building resilience into the economy. Rather than prizing efficiency and just-in-time deliveries, China has invested in redundancies and shock buffers.

China takes energy security seriously. The enormous effort it has made to build low-carbon capacity—solar, wind, and nuclear—has to be understood as part of a broader motivation to make the country dependent on energy sources within its borders. Beijing is trying to mitigate the pain if it ever loses access to sea-lanes that deliver its oil. That is also why, in 2023, China added twenty times more coal-burning capacity than the rest of the world put together. It is serious about addressing issues in climate change, yes. But Beijing is not turning its back on its rich coal reserves. That also explains why China is so enthusiastic about electrifying the auto fleet: It would rather burn domestic coal than Middle East oil to power its cars.

China takes food security seriously as well. Xi Jinping has been known to stand in the middle of a field of wheat while offering a folksy remark: for example, "The bowls of the Chinese people should be filled mostly with Chinese grain." The pandemic and Russia's invasion of Ukraine have made Beijing more conscious of food self-sufficiency. Chinese leaders have always been aware that food shortages have toppled imperial dynasties. And so one of the things that provincial governors are graded on is whether they are self-sufficient in rice and wheat, while mayors of major cities have to make sure that a variety of foods are grown locally. Mayors are graded on the amount of land they dedicate to vegetables and on ensuring that grocery markets are within walking distance for most residents, that there are no food safety scandals, and that prices are stable.

Get on a high-speed train out of Beijing, and you quickly hit farmland. Drive around the outskirts of Shanghai, and you find vast systems of greenhouses growing vegetables. After Xi's remarks, China has in recent years attempted to reclaim salty marshes from the sea and turn idle mines into farmland, even though these are probably not very productive. I don't mind, however, that China is demolishing golf courses, which are environmentally wasteful, and giving that land to farmers.

The cost of this self-sufficiency drive is that a lot of valuable land around cities is tied up for agriculture, in areas that aren't always suited for growing crops. More important, much more of the Chinese workforce is kept rural: Despite its rapid urbanization over the last generation, China still has twice as much of its population living in rural areas as does the United States. The benefit is that during the Covid pandemic, China didn't suffer intense food shortages. The farmland and greenhouses even around the most locked-down cities—Wuhan, Xi'an, and Shanghai—were producing food, but it couldn't be delivered by an overwhelmed logistics system to every resident. Whereas China's food system provided fairly stable production, food insecurity spiked among low-income Americans at the start of the pandemic in 2020. Meat and vegetable production is concentrated in relatively few places. When workers fell ill at slaughterhouses in the Midwest, grocery stores on the East Coast ran out of beef.

Food was not the only item to run short during the pandemic in the United States. Many different items were hard to find: furniture, semiconductors, personal protective equipment. The Chinese government and Chinese companies tend, on average, to maintain greater stockpiles of different goods so they have better resilience. The American corporate dictum is that "inventory is evil." Although having spare capacity hurts various profit measures of Chinese firms, especially its state-owned enterprises, they are better able to leap into action in any crisis. A lot of manufacturing and food capacity is a useful thing to have if there is another pandemic—or a war.

• • •

THE MOST IMPORTANT THING that the engineering state is set up to do is to build manufacturing capacity. Though China faces many headwinds, it is continuing to strengthen its position in a wide range of technologically intensive industries as well as in its military capacity. Even if the United States is able to outclass China in diplomacy, finance, and innovation, the contest between these two great powers is going to be close if the United States can't build anything in the physical world.

The strongest wind in China's sails is the entrenched technological workforce that preserves process knowledge that I wrote about in Chapter 3 on tech power. Though 50 percent of China's economy might be dysfunctional, 5 percent is doing superbly well (an approximation I borrow from Greg Ip at the *Wall Street Journal*). That 5 percent is dangerous for American interests: It is China's manufacturing capability, chiseling away at the American industrial base.

Remember that Chinese companies totally dominate many parts of the clean technology supply chain, especially related to solar and batteries. They're still exporting electric vehicles around the world (though many of those exports are products of foreign companies like Tesla). They've gained ground on all sorts of advanced manufacturing, like consumer drones, industrial robotics, and steel presses. China is still behind on semiconductors and aviation, but it has established supply chains in these areas and is determined to catch up. A lot of the groundwork for China's successes were laid before Xi took office. These buzzing ecosystems of technology production are made up of designers, engineers, and technicians who meet every day to solve problems. Their lives don't necessarily depend on policy developments from either Beijing or Washington, DC.

It's also about people. China has around a hundred million people working in manufacturing. The country's population is declining, yes, but it's important to keep in mind that only a thin slice of the workforce is engaged in technological production. Germany and Japan are mighty exporters with, respectively, eight million and ten million manufacturing workers. A country doesn't need so many people to have a robust semiconductor industry: A few hundred thousand highly trained workers are enough. In 2025, China will graduate more than twice as many PhDs in STEM fields as the United States—and many in American universities are Chinese nationals likely to repatriate.

Making China technologically powerful has become a major priority for Xi's third term. He talked about it at the start of his first term, when he remarked that China's greatest historical problem was its lack of technology. In Xi's telling, China was unable to keep up with modernity, as the Qing empire had rotted from within while besieged by "Western ships and their cannons." Subsequently, his government announced Made in China 2025, a sweeping plan to dominate ten technological industries. In 2023, Beijing announced the creation of a new high-level body: the Central Science and Technology Commission. And the following year, Xi declared that the country must become a "science and technology superpower" by 2035.

"The competition for national strength," goes one commentary from the Ministry of Science and Technology in 2024, "is essentially a contest of scientific and technological innovation, ultimately proving which political system is superior." It is a strange sort of declaration, implying that countries should not be judged by whether they create better economic outcomes, generate greater aesthetic or intellectual flourishing, or produce some more general measure of well-being for the population. In the last Cold War, the United States and the Soviet Union argued over broader measures of success. For a segment of elites in Xi's China—echoing the beliefs of the Industrial Party—who can do better on science and technology determines all.

In a crucial way, the United States accelerated China's progress on science and technology. In his first term, Donald Trump unleashed a trade war against Chinese exporters and a technology war against its leading companies. His administration designated Chinese tech leaders—Huawei, drone-maker DJI, chip leader SMIC—on opaque sanctions lists, which throttled their ability to access American technologies. A few were pushed to the brink of collapse. Concurrently, Trump's Department of Justice subjected scientists (mostly of Chinese heritage) to the tender mercy of the US criminal justice system, usually for charges related to relatively low-level problems implicating research integrity. Joe Biden broadened technology controls, demanding that all advanced chips and chipmaking equipment be approved by the US government before they could be sold to China.

I spent years covering the twists and turns of these technology restrictions. The more it went on, the more I felt that the United States was committed to a strategy of destroying its scientific and industrial establishment—through prosecutions of scientists and cutting off the sales of chipmakers—in order to save it. Rather than realizing its own Sputnik moment, the United States triggered one in China.

China's technology leaders have always bought American chips because they wanted to sell globally competitive products. They ignored Beijing's beseeching to buy from domestic vendors for the simple reason that Chinese technologies were not good enough. But the Trump administration gave China's tech leaders every reason to fear being cut off from American technologies. And so the US government fully aligned those Chinese firms that were previously reluctant to build up the domestic industrial base to Beijing's self-sufficiency agenda. All the money and engineering talent that China's most dynamic tech companies used to send to the United States were now staying at home.

Was it worthwhile to devalue the reliability of American companies, not just to Chinese firms but to companies around the world? So
far, export restrictions haven't dealt a decisive blow to Chinese tech
companies, which have found ways to limp along without full access
to American chips. Even Huawei, which suffered the most intense US
restrictions, is still selling 5G equipment globally and smartphones at
home. Sometimes I think that the United States' tech competition with
China—chaotic policymaking under Trump, porous implementation
under Biden—has ended up in the worst of all worlds. These restrictions have scorched China's most dynamic companies without killing
them, which riles them up to break free of American restrictions.

Meanwhile, Beijing is eagerly funding technology development.
As bank lending to real estate projects has collapsed, funding has
surged toward manufacturers. Beijing partially engineered this scenario.
Though China has lost wealthy and creative types, it has been gaining
scientists. Since 2020, high-profile scientists of Chinese descent have
left the United States, pulled as much by China's generous offers of
research funds as they were pushed by the Trump administration's
investigations of research impropriety. Fewer than 1,000 scientists of
Chinese descent moved from the United States to China in 2010; more
than 2,500 did in 2021. A wave of positive media in China has greeted
the biologists or mathematicians that move from an elite American
university to China. Xi probably doesn't mind trading disgruntled
youths for senior scientists.

Is it possible to do science in a tightening political environment?
A common contention I hear is that China can't innovate because it
"doesn't have free speech."

There's no question that Xi has tightened the country's already
limited space for free speech. Free thought is essential for the human-
ities and the social sciences. But I'm not so sure that it's a necessary
condition for the natural sciences, for very little in chemistry, physics,

mathematics, and engineering is innately political. Plenty of autocratic systems in history have delivered startling technological advances.

German states, for example, have done just that. The nineteenth-century Prussian state combined autocracy with the invention of the modern research university. After Bismarck unified the German states under Prussian rule in Berlin, the country became the pioneer in chemicals—arguably the first science-based industry—as well as in electrical engineering. Nobel Prizes in the sciences continued to be awarded to Nazi Germany while it enlisted its scientists to make *Wunderwaffen* like the world's first ballistic missiles and jet fighters for the war. The Soviet Union provides an even starker example. Its scientific establishment conducted groundbreaking research throughout Stalin's Terror. The state had arrested a remarkable number of scientists, including the chief theoretician of the hydrogen bomb and the head of the Soviet space program. More than one scientist had barely staggered out of Stalin's gulags before doing the work that would win him a Nobel Prize. The Soviets built the atomic bomb under the direction of Lavrentiy Beria, Stalin's odious chief of police. Just as in Nazi Germany, the Soviets kept making scientific advancements during the period of the most intense tyranny.

Modern China is nowhere near as extreme as the police states run by Stalin or Hitler. How is it that science can coexist with autocracy? Mostly, I believe, because the precondition for science is that abundant funds are far more critical to science than free speech, and that is something dictators can deliver.

Perversely, repression might encourage scientists to throw themselves still further into their work rather than paying attention to the rest of the world falling apart around them. I don't believe that autocracy is *good* for science, only that it doesn't guarantee its destruction. China has gotten plenty far on industrial advances—solar power, electric vehicles, robotic arms—in an atmosphere of worsening political repression. Now Xi is shoveling money toward scientists. I've interviewed over two dozen scientists in China, most of whom were trained in the United States, who have told me that it's easier to receive funding in Chinese universities than in American universities. Their money comes without many strings attached, whereas a grant application to the National Science Foundation demands fastidiousness on formatting, endless reporting requirements, and the threat of jail if they don't make a proper disclosure.

I envision China becoming something like a more successful East Germany, a state that combines surveillance and political controls with strong outcomes in science and technology. The Communist Party will not relent on the political atmosphere; meanwhile, it will continue its pursuit of science and technology. Though East Germany was a leader within the Soviet bloc, it was still behind the West, but I expect China to be more successful. Chinese firms will produce high-quality products, perhaps lagging behind global leaders by only a few years and in only a few industries. They'll make chips not powerful enough to fit into the latest iPhone but good enough for electric vehicles and drones, planes not as efficient as the latest from Airbus but good enough to fly between Bangkok and Shanghai.

My focus, for much of the past decade, has concerned China's rejection of advanced technology with American characteristics. And though I would love for China to adopt greater legal protections for people, I'm not sure the technology path it has chosen has been unwise.

Throughout this book, I've avoided calling Xi's regulatory storm a "tech crackdown." While disciplining digital platforms and the real virtual economy with one hand, Beijing has with its other dispensed favor to harder technologies like semiconductors. Xi was trying to reorient technology companies to be less focused on virtual or financial innovation, and for the best and brightest from Tsinghua and Peking Universities to work in strategic industries instead.

Underlying Beijing's actions against digital platforms is a suspicion that tremendously profitable digital companies are not producing value for the rest of society. Entrepreneurial dynamism in online education, social media, or fintech are producing various forms of social harm. The virtual economy, including cryptocurrencies and the metaverse, sucked up too much talent and money. Xi and the rest of the Politburo were discomfited that the cutting edge of the economy seemed to have been driven by the vagaries of investors rather than the interests of the state.

My sense, while I watched the crackdown unfold in China, was that Beijing was trying to avoid the economic structure of the modern United States. Over the past two decades, the major American growth stories have been in Silicon Valley on one coast and Wall Street on the other. Subsequently, both tech and finance have been blamed for many social ills. If there is an era of American innovation that attracts Beijing, it might be the Silicon Valley of the 1960s and 1970s. Chipmakers like Intel were hitting their stride, becoming, in part, major suppliers to the Pentagon and NASA. That was a period when tech companies manufactured stuff, employed big workforces, and minded the state's national security needs.

So Beijing attempted economic surgery. China's leadership wanted dynamism in science-based industries that can patch its strategic deficiencies. In particular, that meant advanced manufacturing industries like semiconductors or clean technologies. It meant that China needs to keep producing and "never deindustrialize." Beijing understands social media sites, like Facebook or TikTok, primarily as freewheeling platforms of expression. They bring little gain in economic productivity while creating huge potential for political unrest. Meanwhile, the Chinese leadership looks more longingly at places like Germany, a country that hasn't developed digital giants but is firmly grounded in manufacturing industries.

In the United States, physics and mathematics PhDs hardly have a chance to consider working in their field before a tech giant or hedge fund picks them up at the sidelines of a conference, flashes them with a humongous pay package, and folds these eager minds into their glamorous embrace. Senior government advisers have more or less stated that Beijing intends to block these temptations. Yao Yang, a dean at Peking University, has remarked with satisfaction that salaries have fallen in the financial industry after regulators imposed a salary cap of $400,000 on the financial sector. Its idea, Yao said, is "to reduce the attractiveness of finance and to increase the development of manufacturing."

The strategy has backfired in a major way. Most notably, it has dampened the animal spirits among entrepreneurs after so many had their businesses crushed. And it's also unlikely that running major technological industries like national security science projects will always produce winners. The Soviet Union, after all, ultimately failed to keep up with the technological frontier set by the West, even though it was doing great science. China has created successful commercial firms in a way that the Soviets never did, though they risk being engulfed by the state. In much the same way, the United States still has a certain degree of manufacturing excellence, represented by firms like Tesla. But that is an outlier. Though Tesla might lead the country toward manufacturing strength once more, it might also be engulfed by the diminishing levels of process knowledge that have dragged down the formerly mighty Boeing and Intel.

Though Chinese firms labor under political restrictions from Beijing and chip restrictions from Washington, DC, they have delivered breakthroughs. DeepSeek, made by a Hangzhou-based company, is one of a handful of frontier AI models, with costs that are a fraction of those demanded by OpenAI's ChatGPT. Chinese AI researchers haven't been laggards. They publish a great number of papers on AI, and its companies have released models that score highly on technical benchmarks. Furthermore, the state is deploying AI, but more for the purposes of censorship, facial recognition, and other means of control.

China has advantages it can bring to bear in artificial intelligence. It's becoming increasingly apparent that American companies are not so much constrained on computing power as they are on electrical power. AI data servers are so energy hungry that Microsoft has attempted to restart the infamous Three Mile Island nuclear plant, and Meta was about to build a data center (running also on nuclear power) until it was halted by the discovery of a rare species of bee near that site. Well, nothing thrills the engineering state like gigantic investments in energy production for industry. What China lacks in technological sophistication, it might make up for in electrical power.

There is also a risk that China misapplies AI. The Chinese system is sometimes overenthusiastic about new technologies or new theoretical ideas. In 1978, one of China's top scientists went abroad to learn about an exciting science called cybernetics and took back home the seeds of an idea that bloomed into the one-child policy. Perhaps the lawyerly society will have the ideological resilience not to be seduced by artificial intelligence, while authoritarian countries wreck themselves by doing so. But it's also possible that Western minds will be broken by AI. In the United States, every shift in mass media—from cable television in the 1990s, the internet in the 2000s, social media in the 2010s, and now AI—has increased discontent between the masses and the elites, as well as between the elites and each other. American society has become much messier than two decades ago, when people were bound by a consensual reality rather than spinning off into different worlds.

It's not clear for which country AI will prove more destabilizing. Fortress China is being protected from the ravages of social media platforms. By putting strict limits on the internet and AI, Xi has built China into a security state able to police vast information flows. The hope from Beijing might be that Americans will be driven mad by the dangerous storms produced by the double whammy of social media plus artificial intelligence. Perhaps these things will magnify the internal divisions of Americans. As more Americans retreat into a digital phantasm, Xi will be shepherding Chinese through the physical world to make babies, make steel, and make semiconductors.

And AI shouldn't distract us from broader American deficiencies. I do not think that outright war between the United States and China is certain to happen. But each side is closely studying the other's military strengths and weaknesses in anticipation of conflict. If it does come to pass, it would be an apocalyptic scenario for the world. War might erupt in the Pacific or elsewhere. As relations between the United States and China become more hostile, the chances of conflict grow. The United States is facing a peer competitor that has four times its population, an economy with considerable dynamic potential, and a manufacturing sector that can substantially outproduce itself and its allies. If China and the United States ever come to blows, they would be entering a conflagration with different strengths. Which would you rather have: software or hardware?

The quantitative disparities between the United States and China are stark. In 2022, China had nearly 1,800 ships under construction, and the United States had 5. US support of Ukraine against Russian aggression also exposed the paltry state of its domestic munition capacity. In two days, Ukraine could fire as many shells as the United States makes in a month. At the very end of the Biden administration, National Security Adviser Jake Sullivan said bluntly that the United States will experience "exhaustion of munition stockpiles very rapidly" if it were ever to face the Chinese military.

China does not lack for munitions. In the case of an emergency, it will be able to scale up production of munitions, just as it has with personal protective equipment, while the United States stumbled on basic things. And I worry that the United States is counting far too much on AI to change the tide. Even if the United States achieves artificial general intelligence, it will need to be able to actually manufacture drones or munitions; algorithms alone will never win a battle. Though the United States has the most sophisticated fighter jets and submarines in the world, it makes precious few of them. The US defense industrial base does not often target efficiency when it distributes production to the jurisdictions of favored members of Congress.

In the modern world, many manufactured products can be refashioned for military purposes. The smartphones we carry around have sensors that would have been military grade a decade ago. The consumer drone is also dual use, which is why Ukrainians and Russians have tried to buy China's DJI drones for the battlefield. That's why industrial capacity should be understood, increasingly, as military capacity. All the drones, smartphones, and batteries that are overwhelmingly produced in China give it an advantage that the United States does not necessarily have.

China's large and adaptive manufacturing base keeps growing. In 2024, the United Nations Industrial Development Organization forecast that China will have 45 percent of the world's industrial capacity by 2030. The United States, Europe, Japan, South Korea, Taiwan, and all other high-income states combined add up to 38 percent of capacity. In a crisis, China has demonstrated a greater track record of expanding manufacturing production than the United States has, so it's not clear whether wartime conditions would change this ratio. Meanwhile, US manufacturing capacity faces greater erosion as China's manufacturers, boosted by subsidies, produce even if they're missing profits. China's industrial might is a strategic advantage that could overwhelm all the rich countries in the world.

. . .

I LIKE TO IMAGINE how much better the world would be if both superpowers could adopt a few of the pathologies of the other. I don't see much danger that Americans could wake up one day with a government that effectively steamrolls every opposition to building big projects, and I don't expect Chinese will encounter a government at last willing to leave them alone. Rather, I hope that China learns to value pluralism while embracing substantive legal protections for individuals and the United States recovers the capability to build for its people.

I don't want to get rid of lawyers. Rather, I want to help lift the engineers (and also their technocratically minded brethren, the economists) back up. Not to raise them onto a pedestal but to elevate them so that there are other voices in the mix. The United States could use fewer lawyers who devote their careers to litigating the life out of government agencies and more lawyers of the dealmaker bent who are interested in working out how to deliver better services. Law professor Nick Bagley concluded his seminal paper on proceduralism (which I referenced in the first chapter) with a polite, but deceptively powerful, proposition that I want to echo: Lawyers should consider whether they could achieve more by stepping out of the way.

It is harder to see how China could move away from engineers. The emperors practiced absolutism a millennium before any European monarchs whiffed the idea. China's civil society has long been weak, with strong family clans, but not made up of the sorts of religious organizations and military aristocracy that produced political contestation in Europe. And ever since the introduction of the imperial examinations in the sixth century by the Sui dynasty, would-be intellectuals have mostly conformed to studying a curriculum set by the emperor. One reason that China lacks a liberal tradition—focused on protection of individual liberties—is that court intellectuals tended not to develop philosophies based on restraining the emperor or his bureaucracy.

China needs lawyers. Or, to be more precise, the ability for people to decline the state's designs on their bodies, their speech, and their minds.

The country doesn't lack regulations or statutes. Xi provides everything for his friends; for his enemies, he has the law. Since he made it a signature priority to impose "rule by law," the country has drowned in laws and regulations. That doesn't mean rule of law as the West might understand it. Xi has rejected the idea of constitutionalism, and the president of the Supreme People's Court has denounced the idea of constitutional democracy as a "false Western ideal." China lacks a real commitment to respecting individual rights. The state allows only limited scope for citizens to challenge government actions, while the Communist Party is off-limits from lawsuits. The judicial system doesn't always publish the records of a case and regardless has plenty of discretion to make legal challenges go its way or go away.

How might change come? Perhaps through ordinary acts of resistance. China's leaders have for millennia tried to impose greater controls on the people. And the people have developed their own strategies for dealing with this control. Though the state wants to see society as an engineering exercise, the reality of China—immediately apparent to those of us who have spent any time there—is that the country is messy. Daily life in China is far more disorderly than the images projected by state media, in which every village is immaculate and where everyone sits with a straight back as they listen to Xi's pronouncements.

Neither is the Communist Party staffed by a far-planning technocracy, nor is it able to squeeze as hard as it wants to achieve national security. People find ways to adapt around the most onerous demands of the engineers. They wield weapons of the weak. When folks see a flurry of senseless rules from the government, they might react with foot dragging, petty noncompliance, feigned ignorance, and arguing back. The system for negotiability is one reason that people have been able to accommodate themselves to engineers.

It would be a better future if the Communist Party could learn some restraint and put a higher value on the individual. Spending time with young people who have *rùn* is a good reminder that the

Politburo isn't representative of the country. The Communist Party will never be convinced that Chinese kids blissed out of their minds on psychedelics represent a hidden asset for the country. What I see in them, as well as among other Chinese people who do their best to deal with the engineering state, is a steady effort to hold one's own against overwhelming odds. It is a hope that the Communist Party might one day let its people flourish by leaving them alone.

Creative youths weren't the first people from China to have *rùn*. Two decades ago, a pair of people in their mid-thirties emigrated from Yunnan. They weren't nearly as hip as the Chiang Mai kids. But my parents left China for many of the same reasons: feeling disappointment in the country's direction and willing to roll the dice on a better life abroad. They carried me, a seven-year-old child, with them on their way to Canada.

CHAPTER 7

LEARNING TO LOVE ENGINEERS

I N MOVING TO THE West, my parents made a wrenching personal decision based on what amounted to a guess about the future. It was an educated guess, grounded in part in deep family history that included a few troubling encounters with the state. But mostly it was about what lay ahead. Where did they and their young child (me, seven years old) have the best chance of living a good life? Which government, and which set of rules, was better for their well-being? Looking at a world they knew, run by engineers, and an alluring but mysterious one, run by lawyers (not that they knew that yet), they had to make a choice, a bet. All these years later, it's not an open-and-shut case that they made the right call.

Both of my parents were born in Kunming, the capital of south-western Yunnan province. Yunnan folks are reputed to be laid back, more eager to sit over tea and chat through the afternoon rather than drive themselves too hard. Not that there's much to drive toward. The city was a backwater when my parents left and remains lackluster today. The government classifies Kunming as a third-tier city, which has the stagnant salaries and limp property values to prove it. When I think about my parents' culture and their upbringing, I am surprised they made the decision to emigrate. We are a very Yunnan family. My mom, Rachel, and my dad, Frank, each have one parent with deep roots in Yunnan and one parent brought there by the war.

My dad's father, my Yeye, was born into Yunnan's most prominent merchant family. The Zhu Family Gardens was Yunnan's largest residence, with gardens so splendid that it would have fit in among the charmed estates of Suzhou. Near the end of the Qing dynasty in the nineteenth century, the family patriarch oversaw a business focused on the mining of tin and copper, expanding (as successful merchants did at the time) into selling tea, distilling spirits, producing silk, and possibly partaking in the opium trade—although my relatives have been a bit vague when I ask them about this point.

Yeye was born in the Zhu Family Gardens a few years after the collapse of the Qing dynasty. There wasn't much of a fortune left by the time he was born. The Zhu family lost its wealth after it kept siding with political losers: with the Qing before local warlords routed imperial forces, then with the Nationalists before their defeat by the Communists. The head of the Zhu family had already been executed for political disloyalty when my grandfather was born. So Yeye wound up in Kunming with his siblings, scattered and poor.

My grandfather had just enough means to be able to get an education. There he met a woman who had also fallen from elite origins. My dad's mother, Nainai, was born in Nanjing, then the country's capital. Her father was one of several secretaries to Chiang Kai-shek, head of the Nationalist Party. Before the Japanese seized Nanjing, the secretary took my infant grandmother and retreated with the rest of Chiang's government to Chongqing. Nainai once told me about one of her early memories, in which people frantically tried to shush her crying lest she attract the attention of Japanese bombers. From Chongqing she went to Kunming, the second capital of the wartime government.

Nainai met Yeye when they both were training as chemical engineers. In the 1960s, her Nationalist family connection disgraced her. The Communist Party sent her to labor in the countryside, and she was unable to see my dad or his brother for six years. When my dad was five, his brother fell sick from eating a poisonous mushroom (a common affliction among fungi-loving Yunnanese). He tried to send a letter to Nainai to alert her, but since he didn't know how to write the character for "mushroom," he drew one. She recounted how my dad's note filled her with confusion and alarm: "Older brother fell sick from eating a (drawing of a mushroom)." Until the end of her life, Nainai cursed Mao for his crazy schemes to break up families.

My mom's side of the family has rural origins. Her father, my Laoye, was born in the northern province of Henan. As a teen, he barely survived the great famine that struck the province in 1942; his two brothers did not. Laoye attended a school administered by successive regimes that wrested control of Henan: first Nationalist, then Japanese, then Nationalist again, until the Communist victory. He developed a great love for books. Since Laoye's family had perished, he enlisted with the troops, and since he had some schooling and literacy, they selected him to become an officer. He joined the Second Field Army, whose commissar was Deng Xiaoping, dispatched to expel Nationalist troops from Sichuan, Guizhou, and Yunnan.

In the early days of the Cultural Revolution, the army split into different factions, each proclaiming themselves to be more fervently devoted to Mao. My grandfather fell into the faction that got itself labeled "rightist" and therefore outside of political favor. The winning faction confined his unit to work at home to produce furniture. He had no idea how to do that but tackled the project with soldierly fortitude. After Mao's death, Laoye saw action once more in his life, when China invaded Vietnam in 1979. Serving as a propaganda officer, he carried out a job—dropping leaflets on Vietnamese troops urging them not to resist—that in retrospect sounds laughable. Battle-hardened Vietnamese troops who repulsed the Americans only years earlier were not going to surrender to a leaflet.

My family says that I resemble this grandfather more than anyone else: a round face, wider eyes, and higher cheekbones. These features could also come from his wife, my Laolao, who is descended from deep Yunnan stock. Rather than being able to trace her heritage for a dozen generations through a prosperous merchant family, the family origins of my mom's mother are cheerily insignificant. They've been black tea farmers in the south of Yunnan for generations. Several ethnic groups are prominent where my grandma is from. It's a bit of a joke in the family—since I look slightly unusual—that I have Tibetan or Wa heritage through her.

Laolao grew up in a family with a slightly bigger plot of land than others. That enabled her to get an education and move to Kunming to become a kindergarten teacher. Life was good until the Communist Party designated her family a minor landlord, condemning her to a bad class background. So she too was sent away to labor in the fields, apart from her three daughters. Most of her family is still farming black tea in southern Yunnan. Every time Laolao's relatives visited Kunming, mostly to visit the city's hospital, they brought along some tea and a local chicken, which she stewed into a wonderful, golden broth.

Soldier, landlord, traitor, capitalist. Each of my grandparents suffered through Mao's political convulsions. Former wealth and a Nationalist background condemned my dad's side of the family. But the military and rural family history on my mom's side didn't produce political favor either. Mao's China was a churning cauldron, in which people's positions bobbed up and drifted down by design: Mao sought continuous revolution. When I spoke to my grandparents about their experiences, only my Nainai was still bitter. The others chuckled about the futility of their lives in the Cultural Revolution, laughing off the times they were separated from my mom and dad. They told me they didn't suffer especially badly. That's true. None of them starved to death or faced the ritualized beatings that destroyed other families.

While their parents were sent away to the countryside, my mom and dad mostly enjoyed themselves. They remember the Cultural Revolution as a good time when they skipped school and did their part to advance communism by chanting slogans and beating drums. My mom and dad were lucky. By virtue of being urban residents and good students, both were later able to attend university. Both were born in the golden era of 1959. They were part of the generation of people going to universities and starting businesses. After high school, my dad went to Guangzhou to study computer science, and my mom studied thermal engineering in Kunming.

By the time they were in college, Deng Xiaoping had started to dismantle the planned economy. My mom began freshman year with four ration coupons for meat per month. She was careful not to use them all up in the first week so that she could have red-braised pork later in the month too. The ration coupon system had mostly disappeared by senior year, and she was able to eat meat when she felt like it.

But socialism didn't dissipate at once. When my parents graduated from college in the mid-1980s, they were caught up in a Deng program that was a throwback to Mao's agenda: Both were part of a teaching corps sent into a small city to be teachers to middle school students. The state dispatched the two of them to the Yunnan city of Dali—which also happens to be where Silvia and I fled in 2022 to escape the Covid lockdowns. It's hard to imagine a better place to be dispatched. They became a couple in the teaching corps numbering three dozen youths. When my parents married, their fellow teachers made up most of the guests. Nobody had much money at that time. The groom and bride treated their wedding guests to dinner and handed each guest a piece of milk candy afterward.

Once they completed their teaching service, my parents returned to Kunming. The state assigned my dad to teach computer programming at the local university. At that time, an undergraduate degree in computing from Guangzhou was sufficient qualification to be a lecturer in Kunming. And the state assigned my mom to work at a coal plant. The job was filthy. Since my mom loved, as her dad did, to read and write, she found herself editing the internal news bulletin at the plant.

My mom was determined to leave the coal plant and do something in journalism. She grew up speaking standard Mandarin, amid army officers who came from all over the country, rather than the local Yunnan dialect. When she applied for a transfer to the news bureau, the provincial broadcaster noticed her clear and resonant Mandarin and hired her to report on the culture and health beat. Eventually, the bureau promoted her to be a radio news anchor and, occasionally, a TV anchor. Whenever she sees me on TV or hears me on a podcast, she comments on how I sound before she tells me her thoughts on anything I've said. The voice is best, she reminds me, if my speech starts from my tummy, while I should project the sound as if it were emerging from my forehead. (That's a tip for all the people hosting podcasts today.)

My parents emigrated after a spell of gloom in China during the 1990s. Yunnan's economic outlook was dim then. The political and economic optimism that people felt over the past decade collapsed with Deng's order to violently suppress student protesters, which then triggered international sanctions. My parents—a few years older than the protesters—felt crushed as they watched the army take control of Beijing. There was plenty of doubt around the country that Deng would succeed in his reform and opening policy. Countries like the United States, Canada, and Australia were beckoning Chinese to immigrate. It wasn't easy for a couple in their mid-thirties with a small child to move. Their most-prized possessions were the stacks of books piled in our small apartment, few of which they would be able to carry.

But when the Canadian government declared them to be high-skilled immigrants and gave them work visas, they decided to depart.

In February 2000, we found ourselves in the suburbs of Toronto. The timing wasn't great. It was my first time realizing that snowfall could be measured in feet and that it can sit for months and turn into increasingly foul ice. Worse, the dotcom bubble had just burst. My dad's programming skills became at once unmarketable. My mom fell from reading the news in Yunnan to taking on odd jobs in Canada, including as a janitor, garment worker, and massage therapist. We moved to Ottawa shortly thereafter so that my dad could study for a master's degree in computer science. I sometimes got up to no good while my dad studied and my mom worked, but I didn't believe them when they threatened to send me to the army. To my surprise, they followed through. To their surprise, I enjoyed being a Royal Canadian Army Cadet. Twice a week after high school, I would go to the drill hall near Parliament Hill to practice map reading, bivouac, and occasionally marksmanship. The person most pleased about all this was my Laoye, happy that I chose the army like he did.

My parents were always stressed about money while I grew up. I was able to do most of the stuff that other kids did, but every so often, I received a brutal reminder of how little money we had. I didn't go to birthday parties because we couldn't afford to buy a gift. My parents brought me to a facility one winter to pick up, to my delight, a bag of toys for Christmas. The gladness soured when other kids told me, not with gentleness, that I must have been poor to qualify for these toys. We never had boots sufficient for trudging around in the awful Ottawa winters. When we ate out, it was at Subway, which charged five dollars for a footlong sub. Now, I feel a slight revulsion when I catch the distinctive whiff of the Subway breads.

When my dad found a job as a software developer in Pennsylvania, we packed up our life in Canada and moved to the suburbs of Philadelphia. Our timing again was poor: Three months after we left Canada, the US stock market began to convulse in response to the 2008 financial crisis. Thankfully, my dad held on to his job. And I went to high school in Bucks County, which is the sort of place that people describe as bucolic. While I was in high school, my dad told me one day that he had no money to send me to college. I didn't doubt him: The US immigration system allowed only him, not my mom, to work at that time. I went to study at the University of Rochester, one of the few places that gave financial aid to Canadian citizens, offering me nearly a full ride. As soon as I began college, I started working to cover my expenses.

Every so often, I wonder about the counterfactual of what would have happened if my parents never departed from Yunnan. They think about it too.

My mom and dad sometimes feel regret. They emigrated just as China's economic boom began in earnest. The country had joined the World Trade Organization, and Deng's reforms really did release the pent-up entrepreneurial energy of the country. If my parents had stayed in Kunming, they would have been allocated housing units by the state. These homes didn't enjoy the precipitous rise in value seen in Shanghai or Shenzhen, but it would have been a tidy sum of money. They would have been near their parents, their siblings, and their friends. And they could have had better careers rather than restarting their lives in a very foreign country.

When my parents wonder what life would have been like if they stayed, they can just take a look at how the rest of their classmates are doing. In China, schoolmates are lifelong friends. Past a certain age, typical socializing takes place inside a banquet room with twenty or so of your classmates, getting drunk and reminiscing. Looking around the banquet table would give them a sense of what they've missed.

A few of their classmates caught the boom, taking advantage of China's two great sources of wealth creation: owning property (or participating in the great wave of construction) or owning a factory (and participating in the great wave of exports). Since my dad went to college in Guangzhou, he knew a number of businesspeople who made their wealth selling furniture or some other consumer goods. They aren't billionaires. But they have been able to buy a home—and sometimes an investment visa—overseas, drive a German-made car, and take leisurely holidays abroad when it suits them.

My parents have no entrepreneurial instincts. So they would have probably been more like the majority of their classmates who earned their living by drawing a salary. They wouldn't earn so much by American standards—$2,000 a month would be considered good—but they would have wealth from owning perhaps two or three homes around Kunming. Liquidating one of them would be enough to send their child abroad for education: the United States if the property were central, Australia or Canada if it were located in the outskirts. A few of these college classmates might be considered lower middle class. Perhaps they had a bad run in business, maybe they pissed off their boss—who decided not to allocate them an apartment—and all they had was their salary.

"Of course, I wish we never left," my mom sometimes said to my dad and me. Her friends at the provincial broadcaster have enjoyed nice careers in radio or TV, retiring at the state-mandated age of fifty-five with a pension. My mom might be hanging out with her sister, a now-retired nurse, who spends her mornings doing tai chi exercises in the park and her evenings with a singing troupe. She would be caring for her elderly father while being driven crazy by her mom. My parents would have the freedom to try out new restaurants and to spend their copious leisure time with family and college friends.

My dad is more circumspect. "Most of our classmates would trade places with us, you know," he counters. Yes, my mom knows.

It took them a long while to make life work in the West. But they achieved a middle-class footing about twenty years after emigrating.

My dad now works in the IT department for an insurance company. And my mom spends her time at home, glad to be free of laborious jobs. Their house in Pennsylvania is filled once more with books, like it was in Yunnan. On weekends they walk the Pennypack Trail or visit parks like the Delaware Water Gap. Since I worried about leaving them alone as an only child, I brought home a dog while I was in college, which gave them joy for years. Going to Costco is a weekend ritual for them like it is for many immigrant families. They've even taken up pickleball. After they naturalized as citizens, both of them cast votes in the 2024 presidential election.

The biggest beneficiary of my parents' emigration is me. I have no idea what I would be like if I had grown up in Kunming rather than Ottawa. My parents tell me that the children of their classmates have mostly not found jobs that give them much meaning, not even the talented folks who made it to Beijing or Shanghai. My three cousins, who are in their twenties, all live at home with my aunts and uncles, because they don't want to spend their meager paychecks on rent. If my parents never emigrated, they still might have been able to send me to an American university. In fact, they would probably have been able to afford it more easily. But I certainly wouldn't have been able to do the sort of work I'm proud of, like writing this book.

It is because I have benefited from their move that I feel somewhat embarrassed. Guilty, even. My parents are materially impoverished relative to most of their friends. In many ways, they're more spiritually impoverished too. They haven't made many friends in suburban Philadelphia. Going anywhere from their housing development requires driving. To reach an Asian grocery store or a decent Sichuan restaurant, they spend two hours on highways driving to and from Princeton, New Jersey. I tell them that it is mostly their fault that they don't have much of a community. They haven't really made an effort. But they lack the context for being more engaged in this American suburban setting where it's difficult to get to know others.

So why do their classmates envy my parents? Because they live a pleasant life without having to deal with the problems that attend the lives of even well-off Chinese citizens. The Chinese middle class is precariously exposed to changes in Beijing's mood. Those in business have to deal with incredible stress, facing down threats from competitors or the local government. They have a gnawing sense that their lives are being shortened by the air they breathe or food they eat. And they feel deep uncertainties about their property values, the future of economic growth, or whether Beijing will visit some sort of disaster upon them or their companies. Life in China is deeply textured and all-embracing. But the intensity of family and social demands can smother, and the embrace can come unbidden, firmly and unavoidably, from the state. For many Chinese, a life in the American suburbs is worthwhile, even if their relationship with the community feels gossamer thin. Families in China still wonder whether they can establish a better life abroad and ask the same questions that my parents asked before they emigrated. Generations of Chinese people have prospered in the United States, in part, I'm sure, because their cultures are so alike. Millions of people look across the ocean and envision their futures, weighing the drawbacks and benefits, the similarities and the differences, asking themselves, Would it be better there?

• • •

MY PARENTS HAVE A resigned contentedness about their lives. I have, however, a wish. For their benefit, I hope that they move to my favorite neighborhood in New York City: Sunset Park.

Walk south of the wealthy Brooklyn neighborhood of Park Slope—where brownstone homes retail for around $4 million—and you'll reach Sunset Park. Its homes are not so handsome as those brownstones. Until the 1960s, Sunset Park was populated with Italian, Norwegian, and Finnish immigrants, who worked in the maritime trades on the nearby waterfront. Now the neighborhood is dominated by newer immigrants. Townhomes occupy the streets and commerce lines its avenues, where doctors and real estate agents advertise themselves in English, Spanish, and Chinese. Latino businesses line Fifth Avenue, while Chinese stores make up Sixth, Seventh, and Eighth. Chicharrón is on display on the western avenues, while roast duck and poached chicken hang on Cantonese rotisseries on the eastern ones. Papaya and plantain are sold on the west side, and durian and melons on the east. Many of the Chinese stores, annoyingly, accept only cash, but their offerings are worth a trip to the ATM. A few of the grocers offer mitten crabs with bright orange roe in the fall, just like you can find in Shanghai.

At the north is Sunset Park itself, whose name graces the rest of the neighborhood, which offers excellent views of Manhattan and New York Harbor. Its most prominent feature is the Sunset Play Center. This facility has one of the eleven swimming pools that Parks Commissioner Robert Moses opened in 1936, featuring his typically bold designs. The bathhouse is a brick building in art deco style, with a lobby made of ceramic tile and bluestone that rises into a rotunda. Around the pool one might find tai chi practitioners doing their routines. At all hours, teens play while families stroll through.

Chinese would recognize something in Robert Moses. He was an American urban planner who built at breakneck speed. Moses held a dozen titles—a few forbiddingly boring (like parks commissioner) and a few that were considerably more tantalizing (chairman of the New York City planning commission and city construction coordinator). He bulldozed urban neighborhoods to make way for great bridges and highways, vast parks upstate, and a giant dam producing power from Niagara Falls, as well as urban amenities that the city desperately craved, like the swimming pool at Sunset Play Center.

My parents haven't responded to my entreaties to help them move to Sunset Park. I recognize that they took a great risk in their lives—moving abroad with a little one in tow—and no longer have the appetite for another big change. But I wish they could have ended up in a more vibrant place than suburban Philly. They shouldn't have to choose between a typical life in China, where politics can overturn lives at any moment, and a typical life in the United States, where the bulk of people inhabit suburban lifestyles that feel kind of dreary.

No, not everyone has to live in the suburbs. But Americans do often have to choose between poorly governed cities or car-dependent suburbia. I wish that there were more spaces like Sunset Park: a relatively affordable neighborhood in a city connected by mass transit that enables people of different cultures to mix. For all of New York City's flaws, it remains one of the few truly urban places in the United States, dense, walkable, with some degree of economic integration. But rather than continue to improve and modernize such places, as Robert Moses tried to do, we have left them in a weird, disconnected state while funneling the country's abundant talent into creating new virtual worlds. Is that the trade Americans want?

By the time my parents and I immigrated to the United States in the 2000s, Moses had long departed the scene. He was not merely dead; he was discredited. The imperative that drove Moses—improving society through large-scale, government-led projects—had gone to the grave with him. My parents don't know who Robert Moses is.

New Yorkers used to celebrate Moses. Then in 1974, Robert Caro published a biography of him titled *The Power Broker*, immortalizing Moses for his big projects and his equally big lapses in judgment. To call this biography monumental would be an understatement: Caro poured painstaking research and literary power into each of its 1,300 pages. Not coincidentally, *The Power Broker* was also one of the books that played a part in the consolidation of the lawyerly society. On par with Rachel Carson's *Silent Spring* and Ralph Nader's *Unsafe at Any Speed*, it taught Americans to fear and loathe engineers.

Robert Moses, let it be said, was neither a lawyer nor an engineer. But as New York's master builder, he was both—and more. *The Power Broker* can be understood by several lists: the list of Moses's official titles; the list of his construction projects, which included bridges, expressways, and New York landmarks; and the list of his mistakes, flaws, and prejudices, which has made his name an oath against physical change. Moses, as Caro demonstrated, was an elitist who bulldozed poor neighborhoods in the service of the middle class. He was arrogant, unwilling to involve anyone else in the interpretation of the public interest, especially not members of the public. And he connived against anyone—poor or powerful—who dared oppose his plans. Though he burned with zeal, he was also burdened with racism and a penchant for petty vengeance.

When it was first published, the book seemed prescient. *The Power Broker* carried a subtitle fit for its time: *Robert Moses and the Fall of New York*. New York was legendary for being awful through the 1970s, facing urban unrest and the threat of bankruptcy. In one of Caro's most compelling chapters, Sunset Park is described as an irredeemable slum. Moses had barely taken a glance at the neighborhood before he thrust an elevated expressway above its busiest commercial street. The expressway, paraphrasing Caro's vivid imagery, tore the heart out of the neighborhood by driving Finnish eateries and Norwegian shops away for the convenience of trucks. Commerce left when the expressway came, leaving a community to crumble.

Stroll around Sunset Park's restaurants that peddle empanadas and noodles today, and its residents would be bewildered to learn that their neighborhood had had its heart torn out. The expressway Moses built is still there. Working-class immigrants are also still there, only now the locals speak Spanish, Cantonese, or Mandarin. The Sunset Play Center that Moses built for the neighborhood stands also, enjoyed by a new generation of families. Rather than fatally collapsing after losing its heart, Sunset Park has endured. It has endured because the neighborhood is bigger than a single construction project. And it has endured because migrants are still seeking a better life in the United States. I would be delighted to settle my parents in Sunset Park. It would bring them closer to the good parts of China that they no longer have access to: a walkable neighborhood where they could find foods they love, a great park where they could practice tai chi or play pickleball, and the option to take the subway around New York City to visit bookstores and cultural events.

The Power Broker has become a dated book. New York may be flawed, but it hasn't fallen. The city walked itself back from the precipice in a way that industrial cities like Detroit, St. Louis, and Cleveland have not. These cities have lost two-thirds of their population since the 1950s, while New York has grown by attracting not only the working class but also the wealthy.

Like many builders, Moses wielded prejudices and made extravagant mistakes. His reign ended at the right moment: Critics like Jane Jacobs and Lewis Mumford called time on his relentless construction projects before he obliterated Lower Manhattan with yet another highway project. But his legacy of physical dynamism has also propelled New York into the global city it is today. Moses thought more deeply about how to attract families into the city than his critics give him credit for. The cultural centers he built have lent their gleam to a city that continues to attract creatives, too.

What New York has lost since the 1960s are updates to its physical environment. The city is still relying on the infrastructure that came to a stop when the reign of Moses ended, which makes me think that there's not much to be gained in stomping on Moses's name still further. New York, and the United States writ large, cannot survive indefinitely on the infrastructure built nearly a century ago. There are always trade-offs and compromises inherent to building large-scale public works, and instead of vilifying the people from our past who made tough choices, we must confront these tough choices ourselves.

The United States has been weakened not only by a procedure-obsessed left—which has become so determined to avoid the errors of Moses that few big works are built at all—but also a thoughtlessly destructive right. I bring up Moses to suggest that the American left needs to rouse itself to deal with the problems of the present day rather than the problems of the previous midcentury.

The American right, I hope, can remember that it is possible to build wonders using the government. In 2025, the tech right celebrates the achievements of Elon Musk, whose Department of Government Efficiency (DOGE) seeks to shred the federal government. No one can dispute that the US government is capable of astonishing inefficiency, but it used to be able to deliver the technologically astonishing too. If the left can reckon with Robert Moses, the right should reckon with Admiral Hyman Rickover—an engineer who improved national security through a large-scale, government-led project.

Better known as the father of the nuclear navy, Rickover launched the USS *Nautilus* in 1954. It was the world's first nuclear-powered submarine, able to travel underwater for weeks (rather than the diesel-powered crafts that could stay underwater for hours), which represented a decisive advantage against the Soviets when it was first unveiled. He was a perfectionist engineer who had the patience to work for decades within the government to see his vision through. What Rickover delivered is a fleet of submarines that remains the pride of the US Navy today. During World War II, industrialists went into government to scale up aircraft and naval production. The US government concentrated resources to accomplish great technological tasks like the Manhattan Project, which produced the bomb, and Apollo Program, which sent humanity to the moon. These kinds of massive technological feats could only be accomplished *through* the government.

I think about Robert Moses and Hyman Rickover not because they were gentle souls. Each had an unseemly lust for power. Both men were idealists with sharp elbows. Both men, as it happens, were also Jewish, experiencing prejudice in institutions meant to be genteel: Yale University for Moses and the US Navy for Rickover. Both were also devoted public servants who spent their entire lives building great works for government pay. Rickover and Moses achieved something we no longer see among public officials. They delivered projects on time and under budget, year after year, while avoiding corruption charges.

There are still plenty of people with tremendous vision and drive in the modern era. Only they are, like Elon Musk, more likely to found tech companies or hedge funds than to work for the public interest. Or departments like DOGE. The Department of Government Efficiency has brought contempt for government, lopping off core institutions and services. Are billionaires like Musk somehow more accountable than America's prior generation of builders? I submit that they are only more obnoxious. The problem with the American right is not its desire to make the government more efficient. Their problem is that they diagnose the causes of inefficiency as a lazy workforce rather than the mountains of procedure that civil servants labor under. DOGE would be more effective if it targeted reductions in process rather than personnel.

The American right, I hope, can remember that the government is capable of building mighty works too. If ambitious people are mostly working in consumer internet companies, then there's little wonder at the disappointment embedded in Peter Thiel's quip: "We wanted flying car, instead we got 140 characters." Shed a tear for the American states: wounded by the ostentatiously destructive tendencies of the right after it has been strangled and dragged down by the left.

• • •

THE ULTIMATE CONTEST BETWEEN China and the United States will not be decided by which country has the biggest factory or the highest corporate valuation. This contest will be won by the country that works best for the people living in it. The United States has deep and enduring advantages over China. But the engineering state has a powerful card to play: It can harness physical dynamism. China has greater manufacturing capabilities, more sophisticated physical infrastructure, a more robust defense industrial base, and more abundant housing. The United States can prove itself the stronger country over the next century if it can hold on to pluralism while building more.

Right now, it is failing. It won't be able to respond to climate change, drive better economic outcomes, or deliver broader measures of social equality if the physical world remains underdeveloped. American governance is stronger if it can demonstrate that it has a political system capable of delivering essential services to its people, including safe public streets, functioning mass transit, and plentiful housing. For various American ideals to be fully realized, the country will need to recover its ethos of building, which I believe will solve most of its economic problems and many of its political problems too.

The United States will be stronger if it can manufacture. If it does not recover manufacturing capacity, the country will continue to be forcibly deindustrialized by China. US global power will be reduced if people around the world find it more attractive to drive Chinese cars, deploy Chinese robots, and fly Chinese planes. The world is more dangerous if Beijing believes that the United States has insufficient ships and munitions to respond to an aggressive act against Taiwan or in the South China Sea. If the two superpowers fight in East Asia, it's not at all clear that the United States will win. America has to build to stave off being overrun commercially or militarily by China.

The United States will be stronger if it builds more homes. American progressives have a slogan that every billionaire is a policy failure. Since common folks are more on my mind, I propose an amendment: Every rise in housing prices is a policy failure. Prosperous places with substantial job creation—especially New York, San Francisco, and Boston—have perversely done the most to block new housing. Overall, half of American renters are considered cost-burdened (meaning that they spend more than 30 percent of their pretax income on rent), and many people who would like to buy a first home cannot afford one. The lack of building new homes has locked people out of cities with good jobs. It is increasing segregation by class and race.

And the United States will be stronger if it can provide better infrastructure. Though New York has mass transit, most of it was built a century ago, such that entering a subway station in Manhattan feels too often like descending into a rotting pit, where one stands amid trash and worrisome leaks, until a deafening metallic screech announces the train. It's not that the city doesn't spend enough on these problems: New York has the honor of hosting five of the six most expensive transit projects in the world. It costs five times as much to build a kilometer of subway in New York City as it does in Paris. If it only cost twice as much, it might be a national tragedy; since it costs five times as much, it is only a statistic. There's no reason that much older European cities should be able to build more cheaply than New York. And the people in charge don't seem to be able to do anything about it.

To the Biden administration's credit, it made a serious attempt to conduct industrial policy and build up US infrastructure. But the pace of building has been terribly slow. In 2021, Congress allocated $42 billion to expand broadband services to rural communities in a plan known as Internet for All. Four years later, not a single home has been connected to this network. Two years after Congress allocated $7.5 billion to build electric vehicle charging stations across the United States, just seven have become operational. The leisurely pace of construction was a political failure for the Democrats: After winning the 2024 election, President Trump will be either able to reap the political benefits of naming many new bridges for himself or cancel some of these projects.

Representative Alexandria Ocasio-Cortez had an early viral moment in 2019 when she lashed out at people holding up the green transition: "The world is going to end in 12 years if we don't address climate change," she said. "And your biggest issue is how are we gonna pay for it?" You can't look at New York's transit projects, where it costs billions to build a mile of subway, and conclude that the city doesn't pay. The problem, rather, is that the government insists on tripping itself up. The ludicrously slow pace of construction in the United States would not be such a big deal if the world wasn't facing, as Ocasio-Cortez points out, a climate crisis. What the United States has lost sight of is that the public might prefer a government that does *something* rather than one that's so exquisite about process. When public works arrive seriously over budget, when the state is barely able to maintain existing infrastructure, when timelines for a new train or a new station can be more than a decade away, we have to question whether the present approach is fit for purpose.

To succeed, the United States doesn't need to adopt China's means of construction. This book has detailed how the engineering state's approach has wrought horrors and is no longer fit for purpose even in China. Rather, the United States can look toward other Western countries, like Spain, Germany, and Japan, which strike a better balance between public consultation and environmental review on the one hand and getting stuff built on the other.

To achieve all of this, I propose—very gently—to unwind the

dominance of lawyers in the United States. That will require us to confront the proceduralism that exists inside government and broader society. And it will require us to renew our faith in government institutions to deliver essential services.

It is difficult, I confess, to pry American political institutions out of the grip of law schools. They have built not just a pipeline funneling graduates into the federal judiciary; ambitious students have cleared paths toward the White House and executive agencies too. It is even harder to change the more fundamental philosophical basis of American law. The United States inherited a common law system typical for anglophone countries, in which judges have much more discretion (relative to legislatures) to shape the law. It is no coincidence that housing and infrastructure costs are astronomically high across the anglosphere, including in the United Kingdom, New Zealand, and Ireland.

The United States will not overcome the lawyerly society by debating the kinds of issues that law students thrill to: the correct ruling on any particular case or the personalities on the Supreme Court. I want to invoke the classic line by professor Grant Gilmore, in a text often assigned to first-year law students: "The worse the society, the more law there will be. In hell there will be nothing but law, and due process will be meticulously observed."

Rather, I want Americans to experience what the previous generation of Chinese have felt: a sense of optimism about the future driven in large part by physical dynamism. Chinese who have experienced the country's blistering economic growth over the past four decades look to the past with pride and to the future with hope. When residents of Chongqing or Shenzhen see a new cityscape unfold before their eyes, they expect the future to keep changing for the better.

When my parents emigrated, China's economy was growing above 10 percent a year; if they had stayed in Kunming, they would have felt like they were living in a new city roughly every seven years, since that's how long it takes for the economy to double. Each time they return to visit their parents, they discover a new, cleaner, better city. Such growth rates are beyond the United States' wildest dreams. They're also no longer in China's reach. But the engineering state continues to build big works because the political economy is fully geared toward it. My parents traded their life in China for the quiet comforts of suburban Philadelphia. It has been good for us, but we all feel that the United States has become distinctly unambitious.

The path forward demands that we reclaim a sense of optimism: an ability to make plans and deliver on them. The United States has to do two things to overcome the lawyerly society.

First, it has to remember that the country has a heritage of engineering. America built beautiful cities full of monumental buildings. Throughout the nineteenth century, it filled these cities with engineering marvels: the world's then-longest suspension bridge connecting Brooklyn and Manhattan (later superseded in length by the Golden Gate Bridge), the world's first skyscrapers in Chicago, subway lines in New York City that ranked with any in Europe. It built bridges, tunnels, highways, and railroads. It demonstrated the technological sublime, like fleets of nuclear-powered submarines as well as vessels that brought humans to walk on the surface of the moon.

Second, the United States needs to elevate a greater diversity of voices among its elites. The most important American virtue is the commitment to pluralism—the ability of diverse cultures to coexist and thrive under equal protection. It means lawyers should be joined by engineers, economists, and other sorts of humanists to make sure that the country is able to work for the many, not only the few.

. . .

CHINA, CLASSIFIED IN 2025 as an "upper middle-income" country by the World Bank, will in a few years cross the "high income" threshold. Beijing will not celebrate that achievement. "No matter how China's economy develops in the future and how its international status improves," Communist Party propaganda organs blared in 2023, "China will always be a developing country."

I find that beautiful.

This declaration is part of a cynical diplomatic effort to convince the poorer countries of the world that China stands for their interests. That's not the appeal for me. Rather, I think it is wise for the country to declare that it is "developing." The United States should do that too. Isn't it better than to be a "developed" one, which implies that you're done, finished, at the end of the road? Leave "developed" status, I say, to Europe's beautiful mausoleum economy.

Over the past forty years, China has resembled the United States at the end of the nineteenth century. Both were feeling their way into superpower status. It was a time for building great works but also a time when scam artists and swindlers abounded, cheating people of their savings for fantastic investment projects. Both were focused on scaling up established technologies rather than doing great new science. Neither country was a great inventor of new products over these periods. Rather, they stole and copied from real scientific innovators: the United Kingdom and Germany at the end of the nineteenth century, the Western world at the start of the twenty-first.

Then the United States exited its Gilded Age. The masses lost their affection, to the extent they ever had any, for the robber barons and their domination of the political system. American progressives launched all sorts of reforms to set the country on a better path. The country harnessed its commitment of transformation to improve its civil service, build new cities throughout its vast territories, and demonstrate that democracies are not militarily weak.

That commitment to transformation is an ideology that both the United States and China share. The United States has a distinctly ideological character as a nation, founded on values and principles rather than heritage; modern China is intent on proving that its historical heritage is glorious. Both countries have an ethos of self-transformation that have become deformed in various ways. For both countries to develop the potential of its people, they have to figure out how to fully express their transformational urge.

Part of the process that drove Deng Xiaoping to embark on reform and opening was his tours to rich countries: A visit to a Texas supermarket, offering so much choice, overwhelmed him; when he heard that an auto worker at Japan's Nissan might be able to produce ninety-four cars a year, while an auto worker in China could produce but one, he realized this was modernity. Now these roles are reversed. It is China that executes on highly complex tasks and Americans who should be looking on with astonished expressions, wondering if they can recover the ability to do such things themselves.

The Communist Party's method of expressing its transformational urge is a top-down effort to organize centralized campaigns of inspiration, which it deployed to achieve communism and, subsequently, economic growth. There would have been every reason to expect Deng to fail when he became China's top leader in 1980. The country had just suffered through the lethal utopian experiments of the Mao years. Deng unleashed the terror of the one-child policy at the same time as his economic reform program. Reform and opening suffered bruising setbacks over the course of the next fifteen years, especially after Deng ordered the army to clear Beijing of protesters in 1989. But then economic growth really did take hold.

The question all of this is leading up to is, Who is better positioned for the future?

Beijing has been taking the future dead seriously for the past four decades. That is why China will not outcompete the United States. The engineering state has delivered great things. But the Communist Party is made up of too many leaders who distrust their own people and have little idea how to appeal to the rest of the world. They will continue to bring literal-minded solutions for their problems, attempting to engineer away their challenges, leaving the situation worse than they found it. Beijing will never be able to draw on the best feature of the United States: Embracing pluralism and individual rights. The Communist Party is too afraid of the Chinese people to give them real agency. Beijing will not recognize that the creatives and entrepreneurs it is chasing into exile are not the enemy. It will not accept that their creative energy could bring as much prestige to China as great public works.

But there are still some things that the United States can learn from the engineering state. Although the creative class wants to *rùn*, the material benefits for most of China's population are widely spread. The reason that consent of the governed is still pretty strong in China is that Chinese have seen their conditions of life improve immeasurably, such that most people have space in their lives to do most of what they want, most of the time. Part of the hopefulness of prior decades has evaporated under Xi, which is another substantial reason that China will not outcompete the United States. But Xi can still count on momentum from China's many strengths to push the engineering state to achieve astounding building feats over the next decade.

The United States has lost its ability not only to build but also, in part, to govern. The procedure-obsessed left and the destructive right have robbed from the people the sense that physical dynamism is desirable. But the United States has pluralistic values, which positions it to better figure out the right solutions.

I've written this book because the very thing that drew my parents to the United States—not lawyers, but pluralism—still provides the potential for course correction. The ultimate reason to be hopeful for the United States is that it can look to its own history to see the path forward. You can see the musculature of the engineering state amid the mighty industrial works scattered all over the country. There's a natural legacy it has to draw on to stage this next act of transformation.

What the United States presently lacks is the urgency to make the hard choices to build. Americans have to trust that society can flourish without empowering lawyers to micromanage everything. The United States should embrace its transformational urge. I hope one day that America can declare itself to be a developing country too. It can demonstrate that the country is able to reform itself, get unstuck from the status quo, and ultimately unlock as much as possible of human potential. "Developing" is a term to embrace with pride.

ACKNOWLEDGMENTS

GAVEKAL DRAGONOMICS WAS THE best possible place to think about China. When I had lunch with Arthur Kroeber in New York one day in 2016, I didn't imagine that his assignment for me to study China's technology developments would plunge me into an adventure between Hong Kong, Beijing, and Shanghai. Arthur is a source of wisdom and good judgment, not only on China, but indeed all things. Andrew Batson taught me how to be an analyst and made me a better writer. Louis Gave ran the company with good cheer and filled its ranks with deeply curious colleagues. Simon Cartledge, friend of the firm, made Hong Kong a more intellectual city. I am glad to have worked with them all.

And the Yale Law School's Paul Tsai China Center was the best possible place to write a book about China. Professor Paul Gewirtz was the most encouraging mentor imaginable. I am immensely grateful to Paul for arranging a perch for me to reflect on China at just the right remove from the neuroses of both Beijing and Washington, DC. The Paul Tsai China Center was packed with wonderful and knowledgeable colleagues. I am fortunate to have had access to the Mac-Millan Center (which generously named me a lecturer), the Jackson School, and the broader community of scholars at Yale, including Arne Westad, Jing Tsu, Dan Mattingly, Paul Kennedy, Zach Liscow, David Schleicher, and more.

I break into hives whenever I hear anyone offer a highly confident view of what Beijing will do. Others of us know better. We were analysts, journalists, executives, and diplomats who were aware that none of us held more than fragmentary knowledge on what's going on in the heads of the leadership. I am grateful to the scores of people in Beijing, Shanghai, Hong Kong, San Francisco, New York, and Washington, DC, with whom I chatted over coffee, lunch, or drinks to engage in that exercise in humility: piecing fragments together.

Toby Mundy had faith in this book before I had a real idea of how to write it. Toby is the most thoughtful and skilled agent one could ask for at every step of the process, from pitching to production. I count my blessings that Toby steered me toward Caroline Adams, my editor at Norton, who bowls me over with her combination of talent, patience, and enthusiasm. Thanks to Pat Wieland, Rebecca Homiski, Kyle Radler, and the entire team at Norton, which is a dream to work with. Leah Paulos of Press Shop brought this book to the attention of many. At Penguin Press, I could count on Casiana Ionita's steadfast support, while Fiona Livesey helped bring the book to a global audience.

This book wouldn't be what it is without the support and companionship of Hugo Lindgren, who elevated my ambitions and enlivened its stories. I prospered from Hugo's regular infusions of writerly confidence. My thanks to Uri Bram, who introduced me both to Toby (whom he called "the best of all agents") and Hugo ("the world's finest editor"). Uri is correct. If my spirit wavered, I could expect Nick Bagley to rouse me with his exuberance. The lawyerly society came together after I listened to Nick's appearance on the *Ezra Klein Show* and over regular lunches in Ann Arbor. I recommend his forthcoming book as well as his services as a wedding officiant.

This book is stronger from manuscript workshops run by Stephen Kotkin of the Hoover History Lab at Stanford, as well as Henry Farrell and Jessica Chen Weiss of Johns Hopkins SAIS. *Breakneck* is written from a perspective that makes most political scientists tart and many historians grumpy. I am full of thanks that scholars nonetheless came together to read my manuscript and offer feedback: Stephen Kotkin, Joseph Torigian, Joseph Ledford, Glenn Tiffert, Graham Webster, Covell Meyskens, Anthony Gregory, Eyck Freymann, Weila Gong, and Ria Roy in Palo Alto; Henry Farrell, Jessica Chen Weiss, Tom Orlik, Todd Tucker, James Palmer, Jeremy Wallace, Eugene Wei, and Steven Teles in DC. Henry Farrell and Eugene Wei are intellectual teddy bears; every time I see them I want to take them into my arms and squeeze.

Many people offered encouragement for this book project before I commenced writing, most of all Tyler Cowen. I have felt unbroken gratitude since I was in college that Tyler has taken an interest in my work. We have continued the conversation in Dali, Taipei, Virginia, and, I hope, many more places to come. Eva Dou, Noah Smith, Ben Thompson, Tracy Alloway, Brad DeLong, Patrick Collison, Ezra Klein, Chris Schroeder, Simon Cartledge, Yiren Lu, Stephen Green, Yanmei Xie, Kevin Kelly, Arjun Narayan, Kevin Kwok, and many others gave me early encouragement for this book. I am tremendously grateful to Chris Miller, Evan Osnos, James Crabtree, Tim Hwang, and Henry Farrell for sharing drafts of their proposals.

I am indebted to Arthur Kroeber, Greg Ip, Nick Bagley, Christian Pfrang, and Ola Rye Malm for reading the entire manuscript. And to those who read it in parts, especially my Shanghai friends who offered their perspectives on the lockdown: Ken Jarrett, Ian Driscoll, Gavin Cross, Mattie Bekink, Victor Bekink, Eric Goldwyn, Christian Shepherd, Teng Bao, Jeff Lonsdale, Chris Delong, Hollis Robbins, Kristina Daugirdas, David Schleicher, John Ryan, Patrick Steigler, Gabriel Crossley, Chris Zheng, and others.

My life in New Haven was simple: I went from the library to the squash courts and then back to the library. On the courts, I am glad to have had Nick Frisch, John Ryan, and Nicholas Bequelin as regular partners, all of us equally endowed with little skill but much enthusiasm. Darius Longarino, Karman Lucero, Changhao Wei, and Jeremy Daum kept the office fun with the occasional board-game night. Paul Gewirtz took me to unbelievably prime seats at the ballet. Soaring Eagle kept things interesting. When I craved some of the stimulations in New York, Dave Petersen and Eugene Wei laid out a spare bed for me. Thank you to everyone who made it fun.

Book writing would have been so much lonelier without the companionship of Silvia Lindtner. She is sensitive not only because she has written a book herself; Silvia is also the wisest and most caring person I know. We have gone on adventures together, we have dealt with grief together, we have debated together, and we have experienced so much joy with each other. I never want our conversation to end.

This book is dedicated to my parents, Frank and Rachel. I respect my mom and dad for having the boldness to *run* before it was a thing, no less with a little one in tow. There's nothing I would change about my childhood. The best lesson I learned as a Royal Canadian Army Cadet was to treat the most difficult things as the most simple things. I want the best for you, which is why I hope you consider the difficult: going out for more exercise, adopting another dog, getting involved with the local community, and, maybe one day, moving to Sunset Park.

NOTES

Chapter 1: Engineers vs. Lawyers

10 **a cost of $36 billion:** Rose Yu, "China's Busiest High-Speed Rail Line Makes a Fast Buck," China Realtime blog, *Wall Street Journal*, updated July 20, 2016.

10 **1.35 billion passenger trips:** "京沪高铁累计旅客发送量约13.5亿人次" [The Beijing-Shanghai high-speed railway has transported approximately 1.35 billion passengers], *People's Daily*, June 24, 2021.

10 **is $128 billion:** Ralph Vartabedian, "New Cost Estimate for California High-Speed Project Puts It Deeper in the Red," *Los Angeles Times*, March 11, 2023.

11 **an extra mountain range:** Ralph Vartabedian, "How California's Bullet Train Went off the Rails," *New York Times*, October 9, 2022.

11 **tout the number of high-paying jobs:** "News Release: Putting Jobs First: California High-Speed Rail Crosses 13,000 Construction Jobs Milestone," California High-Speed Rail Authority, March 19, 2024.

11 **broadly failed to build:** Austan Goolsbee and Chad Syverson, "The Strange and Awful Path of Productivity in the U.S. Construction Sector," Working Paper no. 30845, National Bureau of Economic Research, January 2023, rev. February 2023.

12 **"Sue the bastards!":** Paul Sabin, *Public Citizens: The Attack on Big Government and the Remaking of American Liberalism* (W. W. Norton, 2021), 101.

13　**"American aristocracy":** Alexis de Tocqueville, *Democracy in America* (Library of America, 2004), 309.

14　**In a seminal paper titled:** Nicholas Bagley, "The Procedure Fetish," *Michigan Law Review* 118, no. 3 (2019): 345–401.

15　**"On Wall Street, Lawyers Make More Than Bankers Now":** Cara Lombardo, "On Wall Street, Lawyers Make More Than Bankers Now," *Wall Street Journal*, June 22, 2023.

15　**"Pay for Lawyers Is So High People Are Comparing It to the NBA":** Maureen Farrell and Anupreeta Das, "Pay for Lawyers Is So High People Are Comparing It to the N.B.A.," *New York Times*, July 1, 2024, updated July 2, 2024.

16　**In the words of one 1991 paper:** Kevin M. Murphy, Andrei Shleifer, and Robert W. Vishny, "The Allocation of Talent: Implications for Growth," *Quarterly Journal of Economics* 106, no. 2 (1991): 503–530.

Chapter 2: Building Big

22　**monkey tricksters:** Mark Elvin, *Retreat of the Elephants: An Environmental History of China* (Yale University Press, 2006), 265.

24　**one of every seven guitars:** "Economic Watch: Chinese Guitar-Making Industry Rides on Wave of Belt and Road Initiative," Xinhua, September 21, 2023.

26　**shelters into art exhibitions:** "Chongqing's Air-Raid Shelters Revived as Commercial Venues," Xinhua, January 5, 2024.

27　**ranked fourth among provinces in China by length:** "2022年中国高速公路里程同比增加0.29万公里 其中广东省高速公路里程居于首位" [In 2022, China's highway mileage increased by 2,900 kilometers compared to the previous year, with Guangdong Province ranking first in total highway mileage], Insight and Info, November 22, 2023.

28　**half of Guizhou's children attended high school:** "China National Human Development Report 2013 Sustainable and Liveable Cities: Toward Ecological Civilization" (United Nations Development Programme, 2013), 108.

28　**hike through harrowing mountain paths:** "Long Walk to School through Mountains in SW China," *China Daily*, March 29, 2013.

29　**China had built a second batch of expressways:** Jean-Paul Rodrigue, "Length of the Interstate Highway System and of the Chinese Expressway System, 1959–2021," Geography of Transport Systems, accessed January 29, 2025.

29　**In 1990, there were half a million automobiles:** "Road Map to Easy Ride in Beijing," *China Daily*, September 5, 2003.

29 **in 2024, there were 435 million:** "China Reports Annual Average of 21 Mln New Drivers over Last Two Decades," Xinhua, May 2, 2024.

29 **Shanghai alone moved more containers:** Ria Dutta and Surupasree Sarmmah, "Know the Top 10 Busiest Ports in the US," Container X-Change, September 13, 2023, updated July 12, 2024.

30 **approved construction of eleven new reactors:** "China Approves Record 11 New Nuclear Power Reactors," Bloomberg.com, August 20, 2024.

30 **the state built a new city:** Arthur Kroeber, *China's Economy: What Everyone Needs to Know* (Oxford University Press, 2016), 67.

30 **the average price of an urban apartment fell:** Kroeber, *China's Economy*, 108.

30 **4.4 billion tons of cement:** Vaclav Smil, "How the World Really Works by Vaclav Smil—What Powers Our Economies," *Financial Times*, January 25, 2022.

31 **120 new parks every year:** Hu Min, "Shanghai Aims to Become 'a City in the Parks' by 2025," *Shine*, September 28, 2022.

33 **"Infrastructure investment can be too good":** Michael Pettis, "There's a Cost to Mainland Overinvestment," Carnegie Endowment for International Peace, October 26, 2009.

33 **China's high-speed rail system is economically viable:** Martha Lawrence, Richard Bullock, and Ziming Liu, "China's High-Speed Rail Development," International Development in Focus (World Bank Group, June 6, 2019), 68.

34 **costs balloon:** "2023 Project Update Report," California High-Speed Rail Authority, March 1, 2023, 61.

34 **increasing intellectual and business exchanges:** Xiaokang Wu and Jijun Yang, "High-Speed Railway and Patent Trade in China," *Economic Modelling* 123 (March 2023).

34 **reducing road accidents:** Yue Lu et al., "The Influence of High-Speed Rails on Urban Innovation and the Underlying Mechanism," *PLoS One* 17, no. 3 (March 4, 2022): e0264779.

34 **lowering carbon emissions:** Lawrence, Bullock, and Liu, "China's High-Speed Rail Development," 4.

34 **spared from paying income tax:** International Monetary Fund, Asia and Pacific Department, "A Revenue Mobilization Strategy for China," IMF Staff Country Reports 2024, 050 (2024), A002, accessed January 29, 2025.

35 **Around 10 percent of its GDP:** "Society at a Glance: Asia/Pacific" (OECD, March 19, 2019), 25.

35 **miserly with unemployment insurance:** Nicholas R. Lardy and Tianlei Huang, "China's Weak Social Safety Net Will Dampen Its Economic Recovery," Peterson Institute for International Economics, May 4, 2020.

35 **detained students trying to organize Marxist reading:** Rob Schmitz, "In China, the Communist Party's Latest, Unlikely Target: Young Marxists," NPR, November 21, 2018.

35 **"Even when we have reached":** Xi Jinping, "扎实推动共同富裕" [Promoting common prosperity], *Seeking Truth*, October 15, 2021.

39 **five have less than a dozen flights:** James Mayger, "Next China: Roads to Nowhere," Bloomberg News, July 13, 2023.

39 **the government deleted its own admission:** Rebecca Feng and Cao Li, "A Poor Province in China Splurged on Bridges and Roads: Now It's Facing a Debt Reckoning," *Wall Street Journal*, updated May 21, 2023.

40 **in a primetime documentary:** "反腐专题片揭李再勇'政绩冲动': 三年新增债务1500亿元" [Anti-corruption documentary exposes Li Zaiyong's "achievement-driven impulse": Accumulated 150 billion yuan in new debt over three years], China Central Television, January 8, 2024.

41 **sentenced Li to death:** "Former Senior Guizhou Political Advisor Given Death Sentence with Reprieve," Xinhua, August 13, 2024.

43 **approaching that of Italy's:** Amanda Lee, "China Debt: These 3 Regions Have the Most Daunting Debt Piles—So What Can Be Done about It?" *South China Morning Post*, August 12, 2023.

43 **forcing it to revise down its GDP:** Lucy Hornby, "Top-Tier Chinese City Could See 2017 GDP Revised Down Almost 20%," *Financial Times*, January 11, 2018.

44 **previously unseen levels of productivity:** "China Is Trying to Turn Itself into a Country of 19 Super-Regions," *Economist*, June 23, 2018.

44 **produce around sixty million cars:** Brad Setser, "Will China Take Over the Global Auto Industry?" Follow the Money, Council on Foreign Relations, December 8, 2024.

44 **restarted its production lines:** Yoko Kubota and Clarence Leong, "Why China Keeps Making More Cars Than It Needs," *Wall Street Journal*, updated April 28, 2024.

45 **meager increases to unemployment insurance:** Andrew Batson, "What Would It Have Cost China to Support Household Incomes?" Tangled Woof (blog), August 9, 2020.

45 **experienced record trade in 2022:** Eric Martin and Ana Monteiro, "US-China Goods Trade Hits Record Even as Political Split Widens," Bloomberg News, February 7, 2023, updated February 8, 2023.

46 **As we sipped tea in his office:** Interview by author, October 18, 2021.

47 **onshore stocks dance to their own tune:** Franklin Allen et al., "Dissecting the Long-Term Performance of the Chinese Stock Market" (December 28, 2023), *Journal of Finance*, forthcoming; available at SSRN.

47 **more coal than the rest of the world combined:** Javier Blas, "We're Burning More Coal Than Ever Thanks to China," Bloomberg, December 18, 2024.

49 **"voluntary" resettlement rates:** Yonten Nyima and Emily T. Yeh, "The Construction of Consent for High-Altitude Resettlement in Tibet," *China Quarterly* 254 (March 2023): 1–19.

49 **six times fewer intensive care unit beds:** Chen Chen, "Letting Omicron Loose," Think Global Health, December 16, 2022.

51 **a cool one trillion renminbi:** "Flood Prevention, Disaster Relief Top Priorities in Issuance of 1-Trln-Yuan Gov't Bonds: Official," Xinhua, October 25, 2023.

51 **less growth from each unit of new investment:** "What this means is that investment is becoming less effective at generating growth over time. The marginal product of capital (MPK), the amount of incremental GDP produced by every renminbi of increase in the capital stock, has declined from a peak of 0.4 around to around 0.2 recently." Gavekal, "The Supply-Side Structural Problem," https://research.gavekal.com/article/the-supply-side-structural-problem/.

52 **China spent 13.5 percent of its GDP:** "Estimated Capital Formation and Capital Stock by Economic Sector in China," World Bank Data Catalog, October 21, 2021.

53 **In *Abundance*:** Ezra Klein and Derek Thompson, *Abundance* (Avid Reader, 2025).

53 **waiting on environmental analyses:** "Offshore Wind Market Report: 2023 Edition," US Department of Energy, Wind Energy Technologies Office, August 24, 2023.

53 **added 6 gigawatts:** "US Wind Energy Monitor: 2023 Year in Review," Wood Mackenzie and American Clean Power, March 28, 2024.

53 **China added 76:** Kostantsa Rangelova, "2023's Record Solar Surge Explained in Six Charts," Ember, May 30, 2024.

53 **four times more than the rest of the G-7:** Rangelova, "2023's Record Solar Surge Explained."

53 **Ezra Klein of the *New York Times*:** "Ezra Klein Interviews Adam Tooze," *New York Times*, September 17, 2021.

Chapter 3: Tech Power

57 **Its population soared:** "深圳市第七次全国人口普查公报[1]（第一号）——全市常住人口情况," [2020 Shenzhen Seventh National Population Census Bulletin [1] (No. 1)—City Permanent Resident Population Status], Statistics Bureau of Shenzhen Municipality, May 17, 2021.

60 **A Chinese report from 2009:** Xiaoyi Wang, "Foxconn Longhua Technology Park Tour: Entering Terry Gou's Forbidden City (4)," *NetEase Science and Technology Report*, September 30, 2009.

60 **a cool million workers:** Employees, "Corporate Social Responsibility Report, 2020," Hon Hai Precision Industry Co. Ltd., 2021.

60 **his own golf cart:** Wang, "Foxconn Longhua Technology Park Tour."

60 **offering twenty-five majors:** Rob Schmitz, "Foxconn's Newest Product: A College Degree," Marketplace, April 26, 2019.

61 **three million square meters:** Frederik Balfour and Tim Culpan, "The Man Who Makes Your iPhone," Bloomberg News, September 9, 2010.

62 **cruel teasing from her more successful colleagues:** "四川承诺帮富士康招工: 公务员被迫进厂'顶工'"[Sichuan promises to help Foxconn recruit workers: Civil servants forced to work in the factory], Taiwan.cn, April 29, 2012.

62 **Henan officials "borrowed" workers:** Eva Dou, "How the iPhone Built a City in China," *Wall Street Journal*, July 3, 2017.

62 **"interns" who assembled iPhones:** Yuan Yang, "Foxconn Stops Illegal Overtime by School-Age Interns," *Financial Times*, November 22, 2017.

62 **recruited retired People's Liberation Army personnel:** Chang Che and John Liu, "An iPhone Factory Needs Workers: The Chinese Government Wants to Help," *New York Times*, November 18, 2022.

62 **"I need to build a city":** Helen Wang, phone interview by the author, July 16, 2024.

63 **one-eighth of rural buildings:** Bernard Chang, Jeffrey Inaba, Rem Koolhaas, and Sze Tsung Leong, *Great Leap Forward: Harvard Design School Project on the City* (Taschen, 2001), 245.

65 **"[Apple products are] not designed":** Glenn Leibowitz, "Apple CEO Tim Cook: This Is the No. 1 Reason We Make iPhones in China (It's Not What You Think)," *Inc.*, December 17, 2017.

65 **In 2019, United Airlines made:** Mikey Campbell, "Apple Spends $150M a Year on United Flights, Shanghai Is No. 1 Destination," *Apple Insider*, January 11, 2019.

66 **A 2012 story:** Charles DuHigg and Keith Bradsher, "How the U.S. Lost out on iPhone Work," *New York Times*, January 21, 2012.

66 **"In the US, you could have":** Leibowitz, "Apple CEO Tim Cook."

66 **One of the former Apple engineers:** Phone interview with author, April 18, 2024.

67 **"If you have a gas leak":** Phone interview with author, March 29, 2024.

67 **"peace dividends of the smartphone wars":** Ben Pauker, "Epiphanies from Chris Anderson," *Foreign Policy*, November 20, 2024.

68 **supply chain grew more "red":** Phone interview with author, November 24, 2021.

69 **25 percent of the final value:** Yuqing Xing and Shaopeng Huang, "Value Captured by China in the Smartphone GVC: A Tale of Three Smartphone Handsets," *Structural Change and Economic Dynamics* 58, no. C (2021): 256–266.

72 **One of my favorite books about China:** Simon Leys, *The Hall of Uselessness: Collected Essays* (New York Review Books, 2014), 242.

72 **Ise Grand Shrine (or Ise Jingu) in Japan:** Brian Potter, "Ise Jingu and the Pyramid of Enabling Technologies," Scope of Work, February 3, 2023.

73 **"I will leave these duties to you":** Junko Edahiro, "Rebuilding Every 20 Years Renders Sanctuaries Eternal: The Sengu Ceremony at Jingu Shrine in Ise," JFS Japan for Sustainability, September 10, 2013.

74 **spent $69 million to relearn:** "Nuclear Weapons: NNSA and DOD Need to More Effectively Manage the Stockpile Life Extension Program," Government Accountability Office, March 2009.

74 **Andy Grove, the legendary former CEO of Intel:** Andy Grove, "How America Can Create Jobs," Bloomberg News, July 1, 2010.

76 **fell to just eleven million:** "All Employees, Manufacturing [MANEMP]," Federal Reserve Bank of Saint Louis, January 10, 2025.

76 **Michael Boskin, quipped:** Michael Schrage, "Potato Chips vs. Computer Chips: High Technology Any Way You Slice It," *Washington Post*, January 21, 1993.

76 **nearly $1 trillion a year on defense:** "Chart Pack: Defense Spending," Peter G. Peterson Foundation, February 3, 2025.

77 **one to three years behind schedule:** Megan Eckstein, "US Navy Ship Programs Face Years-Long Delays Amid Labor, Supply Woes," *Defense News*, April 3, 2024.

77 **"excruciating":** David Gelles, James B. Stewart, Jessica Silver-Greenberg, and Kate Kelly, "Elon Musk Details 'Excruciating' Personal Toll of Tesla Turmoil," *New York Times*, August 17, 2018.

79 **BYD saw its sales decline:** "Chinese EV Maker BYD Profit down 42% in 2019, Under Pressure from Subsidy Cut," Reuters, April 21, 2020.

79 **Tesla produced the "catfish effect":** Zhao Xinyue and Bo Yuan, "从'年产过千万'看'特斯拉效应'" [Looking at the "Tesla effect" from "annual production of over ten million"], *People's Daily Online*, November 16, 2024.

79 **"Flying with phoenixes":** Yang Jie, "Tim Cook Can't Make iPhones without This Chinese Company and Its CEO," *Wall Street Journal*, updated October 23, 2023.

81 **Economic studies have shown:** Lee G. Branstetter, Li Guangwei, and Ren Mengjia, "Picking Winners? Government Subsidies and Firm

Productivity in China," Working Paper no. 30699, National Bureau of Economic Research, December 2022.

83 **Apple's most recent supplier report:** Guangdong has seventy-two, the United States has twenty-six, India has fourteen, Vietnam has thirty-five. "Apple Supplier List 2023," Apple Inc., 2024.

84 **"The real economy is the foundation":** Zhang Weifu and Hu Yabei, "实体经济: 经济发展的着力点和支撑点" [Real economy: The focus and support point of economic development], EOL, July 13, 2023.

84 **"fictitious" economy:** Hu Yabei and Zhang Weicun, "形成新增长引擎 将着力点放在实体经济上" [Forming a new growth engine rests on the real economy], *Nanjing Daily*, July 19, 2017.

84 **the 419 industrial product categories:** "构建以先进制造业为骨干的现代化产业体系 ——访工业和信息化部党组书记、部长金壮龙" [Build a modern industrial system with advanced manufacturing as the backbone: Interview with Jin Zhuanglong, secretary of the Party Leadership Group and minister of the Ministry of Industry and Information Technology], *People's Daily*, January 10, 2024.

84 **Batson has furthermore detected:** Andrew Batson, "China Wants Those Low-End Industries After All," Tangled Woof (blog), October 4, 2023.

84 **"the real economy":** Xi Jinping, "国家中长期经济社会发展战略若干重大问题" [Several major issues in the national medium and long-term economic and social development strategy], *Seeking Truth*, October 31, 2020.

86 **A few of these works:** Wang Xiaodong, "A Study of the 'Industrial Party' and the 'Sentimental Party,'" Center for Strategic Translation, January 1, 2011, trans. October 2023.

87 **His 2005 book:** Dylan Levi King, "China's Exit to Year Zero," *Palladium*, April 9, 2021; Zhong Qing, 刷盘子还是读书 [*Wash Dishes or Study?*] (Contemporary China Press, 2005).

87 **Ma Qianzu:** Pen name of Ren Chonghao.

88 **China is not yet ready to sever ties:** Vivian Wang, "A Godfather of Chinese Nationalism Has Second Thoughts," *New York Times*, October 27, 2022.

89 **science fiction trilogy by Liu Cixin:** Liu Cixin, *The Three-Body Problem* (Tor Publishing, 2014).

89 **"blood-drenched pyramids":** Liu, *Three-Body Problem*, 229.

90 **China's goods exports:** Eric Martin and Ana Monteiro, "US-China Goods Trade Hits Record Even as Political Split Widens," Bloomberg News, February 7, 2023, updated February 8, 2023.

92 **a state media headline captured:** Zhao Yimeng, "Huaqiangbei Traders Computer Chips for Lipsticks," *China Daily*, updated February 3, 2021.

Chapter 4: One Child

96 **enable "hundreds of millions":** "习近平同全国妇联新一届领导班子集体谈话" [Xi Jinping addresses the new leadership of the All-China Women's Federation], CCTV, October 31, 2013.

96 **The *Economist*'s headline:** "China Wants Women to Stay Home and Bear Children," *Economist*, November 9, 2023.

96 **Six million Chinese married:** Thomas Hale, Wang Xueqiao, Tina Hu, and Wenjie Ding, "China's Marriage Problem: Fewer Young People, and Fewer Weddings," *Financial Times*, February 13, 2025.

96 **average of 1.0 children:** Jacob Funk Kirkegaard, "China's Population Decline Is Getting Close to Irreversible," Peterson Institute for International Economics, January 18, 2024.

98 **Mao boasted to Jawaharlal Nehru:** "Minutes of Chairman Mao Zedong's Second Meeting with Nehru," Wilson Center Digital Archive, October 23, 1954.

98 **lose half of the population:** Nikita Khrushchev and Strobe Talbott, *Khrushchev Remembers: The Last Testament* (Bantam, 1976), 255.

98 **"It is a very good thing":** *Selected Works of Mao Tse-Tung* (Foreign Language Press, Peking, 1967), 453.

98 **"Even if China's population multiplies":** *Selected Works of Mao Tse-tung*, Vol. 4 (Foreign Languages Press, 1961).

102 **Song later wrote that he grew "extremely excited":** Song Jian, "Systems Science and China's Economic Reforms," *IFAC Proceedings* 18, no. 9 (August 1985): 1–7.

102 **two determinations:** Susan Greenhalgh, *Just One Child: Science and Policy in Deng's China* (University of California Press, 2008), 161.

102 **four billion by 2080:** Song Jian, Tian Xueyuan, Li Guangyuan, and Yu Jingyuan, "关于我国人口发展目标问题" [On China's population development goals], *People's Daily*, March 7, 1980.

104 **aid of an abacus:** Greenhalgh, *Just One Child*, 204.

104 **handheld calculator:** Greenhalgh, *Just One Child*, 175.

105 **if he were ever attacked:** Song Jian, "Systems Science."

105 **In a 1988 book:** Joel E. Cohen, "Review of *Population System Control*," *SIAM Review* 32, no. 3 (1990): 494–500.

105 **According to Greenhalgh:** Greenhalgh, *Just One Child*, 294.

105 **open letter of 1,600 words:** "关于控制我国人口增长问题致全体共产党员、共青团员的公开信" [Open letter to all Communist Party members and Communist Youth League members regarding the control of population growth in our country], *People's Daily*, September 25, 1980.

106 **a woman needed up to twelve documents:** Thomas Scharping, *Birth Control in China, 1949–2000* (Routledge, 2002), 95.

107 **rural fertility was closer to 2.5:** Tyrene White, *China's Longest Campaign: Birth Planning in the People's Republic, 1949–2005* (Cornell University Press, 2018), 73.

107 **prohibited from returning home:** Steven W. Mosher, *Broken Earth* (Free Press, 1983), 224.

107 **hauled before mass rallies:** Christopher S. Wren, "China's Birth Goals Meet Regional Resistance," *New York Times*, May 15, 1982.

108 **"Any method that reduces fertility":** Susan Greenhalgh and Edwin A. Winckler, *Governing China's Population: From Leninist to Neoliberal Biopolitics* (Stanford University Press, 2005), 225.

108 **"Take all measures":** US Congress, House, China: Human Rights Violations and Coercion in One-Child Policy Enforcement: Hearing before the Committee on International Relations, House of Representatives, 108th Cong., 2nd sess., December 14, 2004.

108 **seized cattle and other livestock:** *China Report: Political, Sociological, and Military Affairs*, Foreign Broadcast Information Service, 1986.

108 **they induced an early birth:** Nicholas Kristof, "China's Crackdown on Births: A Stunning, and Harsh, Success," *New York Times*, April 25, 1993.

109 **injected formaldehyde into a baby's head:** Michael Weisskopf, "One Couple, One Child: Second of Three Articles Abortion Policy Tears at China's Society," *Washington Post*, January 7, 1985.

109 **In reports now censored:** Anne Henochowicz, "Translation: The Hundred Childless Days," *China Digital Times*, May 4, 2021.

109 **deputy head of the provincial committee:** "中共山东省委关于增补山东省关心下一代工作委员会副主任的通知" [Notice of the Shandong Provincial Committee of the Communist Party of China on appointment of the deputy director of the Shandong Provincial Working Committee for Caring for the Next Generation], Shandong Xiehe University, April 19, 2017.

109 **advocated for universal sterilization:** White, *China's Longest Campaign*, 159.

110 **sterilizing mothers immediately:** Scharping, *Birth Control in China*, 109.

110 **China's health ministry statistics show:** Scharping, *Birth Control in China*, 112.

110 **Weisskopf wrote:** Weisskopf, "One Couple, One Child."

111 **weapons of the weak:** James C. Scott, *Weapons of the Weak: Everyday Forms of Peasant Resistance* (Yale University Press, 1985).

111 **essentially second- or third-class citizens:** Kay Ann Johnson, *China's Hidden Children: Abandonment, Adoption, and the Human Costs of the One-Child Policy* (University of Chicago Press, 2017), 92.

111 **the writer Peter Hessler:** Peter Hessler, *Other Rivers: A Chinese Education* (Penguin, 2024), 52.

111 **a phobia of fire:** *China Report: Political, Sociological, and Military Affairs*, FBIS.

112 **special insurance scheme:** Scharping, *Birth Control in China*, 226.

112 **Only half had completed high school:** Scharping, *Birth Control in China*, 186.

112 **In three separate state surveys:** Scharping, *Birth Control in China*, 189.

113 **"At present, the phenomena of butchering":** Li Jianguo and Zhang Xiaoying, "Infanticide in China," *New York Times*, April 11, 1983.

113 **demographers estimate:** Nie Jingbao, "Non-Medical Sex-Selective Abortion in China: Ethical and Public Policy Issues in the Context of 40 Million Missing Females," *British Medical Bulletin* 98, no. 1 (June 2011): 7–20.

113 **"I should never have been born":** Johnson, *China's Hidden Children*, 92.

114 **bust of a large baby-trafficking ring:** "Court Convicts 52 of Baby-Trafficking in China," *New York Times*, July 24, 2004.

114 **Police raids to rescue trafficked children:** Sharon Lafraniere, "Chinese Officials Seized and Sold Babies, Parents Say," *New York Times*, August 5, 2011.

114 **Johnson recounted:** Johnson, *China's Hidden Children*, 132.

114 **one American on an adoption trip:** Bruce Porter, "I Met My Daughter at the Wuhan Foundling Hospital," *New York Times*, April 11, 1993.

115 **Officers snatched at least sixteen children:** Hannah Beech, "Have Foreigners Unwittingly Adopted Victims of Baby-Selling in China?" *Time*, May 11, 2011.

115 **Longhui residents:** Lafraniere, "Chinese Officials Seized and Sold Babies, Parents Say."

115 **"openly killing people for years":** Evan Osnos, "Abortion and Politics in China," *New Yorker*, June 15, 2012.

116 **It collected $200 billion in fines:** "我国累计收社会抚养费1.5万亿 每年200亿罚款去向不明" [China has collected 1.5 trillion yuan in social maintenance fees: Annual 20 billion yuan in fines unaccounted for], Xinhua, December 9, 2014.

116 **sterilized 108 million women:** *2021 China Health Statistics Yearbook*, National Health Commission (China), May 17, 2023.

116 **more than 150,000 children had been sent:** Austin Ramzy and Liyan Qi, "China's One-Child Policy Sent Thousands of Adoptees Overseas: That Era Is Over," *Wall Street Journal*, September 5, 2024.

117 **"Chinese peasants, your name is misery":** Huang Yasheng, *Capitalism with Chinese Characteristics: Entrepreneurship and the State* (Cambridge University Press, 2010), 139.

119 **viewed rural folks as a variable:** Greenhalgh, *Just One Child*, 179.

119 **"The size of a family":** Weisskopf, "One Couple, One Child."

120 **a trio of demographers offered:** Feng Wang, Baochang Gu, and Yong Cai, "The End of China's One-Child Policy," Brookings Commentary, March 30, 2016.

121 **the fertility rate had already fallen:** "New Research Helps Explain Why China's Low Birth Rates Are Stuck," *Economist*, June 1, 2023.

121 **prevented four hundred million births:** "计划生育40余年 我国少生4亿多人" [Family planning has reduced the birth rate by more than 400 million over the past 40 years], *Gansu Daily*, November 12, 2013.

121 **same sort of linear assumptions:** Martin King Whyte, Feng Wang, and Yong Cai, "Challenging Myths about China's One-Child Policy," *China Journal* 74 (July 2015): 144–159.

122 **former propaganda slogans read:** "只生一个好, 政府帮养老"; "三个孩子就是好不用国家来养老."

122 **In an ambitious article:** Song Jian, "Systems Engineering and the New Technological Revolution," *People's Daily*, September 5, 1984.

122 **1,400 years more ancient:** "Song Jian: A Leading Scientist," *China Daily*, updated January 25, 2011.

123 **Maternity wards are starting to shut down:** Zhao Jinzhao, Pan Rui, and Wang Xintong, "In Depth: Maternity Wards Are Latest Victim of China's Falling Birthrate," *Caixin Global*, June 19, 2024.

123 **adult diapers are expected to outsell:** Edward White and Leo Lewis, "Nappy Manufacturers Shift Focus in China from Infants to Elderly," *Financial Times*, November 28, 2021.

123 **fourteen years before China's:** "China's High-Stakes Struggle to Defy Demographic Disaster," *Economist*, April 9, 2024.

124 **In the 2023 meeting:** "十四届全国人大三次会议举行第二次全体会议" [China's national legislature holds 2nd plenary meeting of annual session], www.gov.cn.

124 **"Do Leftover Women Really Deserve Our Sympathy?":** Leta Hong Fincher, *Leftover Women: The Resurgence of Gender Inequality in China*, 1st ed. (Zed Books, 2014), 3.

125 **discover marital infidelity:** Hong Fincher, *Leftover Women*, 20.

125 **One former employee of the Women's Federation:** Shen Lu and Liyan Qi, "China Is Pressing Women to Have More Babies: Many Are Saying No," *Wall Street Journal*, January 2, 2024.

125 **a tax on the childless:** Charlotte Gao, "To Encourage More Births, Chinese Specialists Propose Birth Fund, Childless Tax," *Diplomat*, August 17, 2018.

125 **an unsigned commentary:** "中国报道网时评: 落实三孩政策 党员干部应见行动" [China report network commentary: Party members and cadres should take action to implement the three-child policy], China Reports Network, December 9, 2021.

126 **I think Heinrich Himmler:** Quoted in Johann Chapoutot, *The Law of Blood: Thinking and Acting as a Nazi* (Harvard University Press, 2018), 343.

126 **Chinese general social survey of 2021:** "China's Low-Fertility Trap," *Economist*, March 21, 2024.

126 **women reacted with incredulity:** Ji Siqi, "China Population: County Sparks Uproar by Telling 'Leftover' Women to Marry Unemployed Men," *South China Morning Post*, January 28, 2022.

126 **a rate that fell to 40 percent:** Linda Bollivar, "Ethan Michelson Highlights Gender Injustice in China's Divorce Courts," Hamilton Lugar School, Indiana University, Bloomington, July 13, 2023.

126 **six hundred hospitals officially authorized:** Farah Master, "China Launches Campaign to Crack Down on Illegal Fertility Treatments," Reuters, July 11, 2023.

127 **One woman posted:** "你被街道办催孕了吗?" [Has the neighborhood committee urged you to get pregnant?], *China Digital Times*, October 21, 2024.

127 **may become a threat to public safety:** Phoebe Zhang, "Love and Marriage: China's Dali Bai Region Pledges to Help Its 33,000 Bachelors Find Wives," *South China Morning Post*, August 24, 2024.

Chapter 5: Zero-Covid

130 **residents could disregard Chinese law:** "Foreign Concessions in China," Wikipedia, May 21, 2025.

131 **also full of opium dens:** Frederic Wakeman, "Licensing Leisure: The Chinese Nationalists' Attempt to Regulate Shanghai, 1927–49," *Journal of Asian Studies* 54, no. 1 (1995): 24.

135 **She miscarried outside the hospital:** Christian Shepherd, "Tales of Anguish Emerge from China's Locked-Down Xian, as Hospital Staffers Are Fired over Woman's Treatment," *Washington Post*, updated January 6, 2022.

137 **a headline on March 24:** Xing Yi, "Shanghai Has No Plans for City Lockdown," *China Daily*, updated March 24, 2022.

137 **"too important to lock down":** "上海新增感染者持续高位，为何不能'封城'? 回应来了!" [The number of Covid infections continues to rise in Shanghai, but here's why there cannot be a lockdown], *People's Daily Online*, March 26, 2022.

137 **Shanghai announced:** "关于做好全市新一轮核酸筛查工作的通告" [Notice regarding the implementation of a new round of city-wide nucleic acid screening tests], Shanghai Municipal Health Commission, March 28, 2022.

137 **ever so softly worded:** Zhao Yusha, Chen Qingqing, and Qi Xijia, "Shanghai Enters Partial 'Pause,' Vows Sufficient Daily Supplies," *Global Times*, March 28, 2022.

138 **barking voice would yell at him:** Aaron Mak, "All the Invasive Ways China Is Using Drones to Address the Coronavirus," *Slate*, February 4, 2020.

138 **"The drone will try to dissuade":** Ann Cao, "Shanghai's Yangpu District Deploys Drones to Detect Violations of Covid-19 Rules, Leading to Complaints," *South China Morning Post*, August 15, 2022.

138 **"Repress your soul's yearning for freedom":** Rob Schmitz, "When This Shanghai Building Went into COVID Lockdown, My Wechat Message Group Blew Up," National Public Radio, April 30, 2022.

140 **a driver holding up bottles of his own excrement:** Luo Chunhao and Yan Yucheng, "千万货车司机困在疫情里" [Millions of truck drivers stranded due to the pandemic], *China Digital Times*, April 11, 2022.

141 **Celebrities complained online:** Henry Lau, "How 6 Hong Kong Celebrities Survived the Shanghai Lockdown, from Carina Lau and Gigi Lai's Covid-19 Testing Pics on Weibo, to Rain Lee's Cheerful Live-Streams," *South China Morning Post*, April 20, 2022.

141 **One of the country's top venture capitalists:** "Shanghai's Locked-Down Elite Are Joining Hunt for Groceries," Bloomberg News, April 8, 2022.

144 **A producer from CNN:** Serenitie Wang, "Shanghai Surprise: How I Survived 70 Days Confinement in the World's Strictest Covid Lockdown," CNN, June 17, 2022.

144 **chasing down a corgi:** Jessie Yeung, "A Covid Worker Beat a Dog to Death in Shanghai after Its Owner Tested Positive," CNN, April 8, 2022.

144 **One woman told a reporter:** Stella Yifan Xie and Liyan Qi, "In Shanghai, Strict Covid Rules Separate Children from Parents," *Wall Street Journal*, April 3, 2022.

144 **separating babies and infants from parents:** Brenda Goh and Engen Tham, "Shanghai Separates COVID-Positive Children from Parents in Virus Fight," Reuters, April 2, 2022.

145 **After an outcry online:** Helen Davidson, "China: Editorial Says Communist Party Members Must Have Three Children," *The Guardian*, December 9, 2021.

145 **denied treatment at the hospital where she worked:** Natasha Khan, "Shanghai Nurse's Death Fuels Skepticism over Cost of China's Covid-19 Measures," *Wall Street Journal*, March 25, 2022.

146 **confronting a police officer:** Mandy Zuo, "Shanghai Lockdown: Residents in Fear of False-Positive Covid-19 Tests after Couple Who Tested Negative Hauled Off to Quarantine," *South China Morning Post*, April 11, 2022.

148 **state media praised performers:** "1月21日，2020年湖北省春节团拜会文艺演出在洪山礼堂圆满举办" [On January 21, the 2020 Hubei Province Spring Festival Gala Cultural Performance was successfully held at Hongshan Hall], NetEase News, January 23, 2020.

149 **one-line snapshots of personal stories:** Alexander Boyd, "Translation: Weibo User Sentenced to Six Months over Wuhan Poem," *China Digital Times*, February 24, 2021.

150 **In the 1990s, Henan province suffered:** Chris Buckley, "Dr. Gao Yaojie, Who Exposed AIDS Epidemic in Rural China, Dies at 95," *New York Times*, December 10, 2023.

150 **George Gao, offered a boast:** "中国疾控中心主任高福: 不应对中国疫苗失去信心" [China CDC director George Gao: We should not lose confidence in Chinese vaccines], *Beijing Youth News*, March 5, 2019.

151 *People's Daily* **declared:** "抗疫斗争彰显中国制度优势" [Pandemic controls demonstrate the superiority of China's political system], *People's Daily*, September 17, 2020.

153 **To access the showers at Shanghai University:** Zhang Wanqing, "In Locked-Down Shanghai, Students Adapt to a Surreal New Normal," *Sixth Tone*, April 30, 2022.

153 **Thirty thousand visitors were trapped:** Alex Binley, "Shanghai Disney: Visitors Unable to Leave without Negative Covid Test as Park Shuts," BBC, October 31, 2022.

154 **a thousand people locked up for days:** "上海基层愤书抗疫三建议, 呼吁反对官僚主义" [Shanghai grassroots officials write angry letter with three suggestions on pandemic control, call for opposition to bureaucracy], Yibao China, April 2, 2022.

156 **James C. Scott has written:** James C. Scott, *The Art of Not Being Governed: An Anarchist History of Upland Southeast Asia* (Yale University Press, 2009), 8.

158 **One of the most-shared essays:** Changhao Wei, " 'State of Emergency' and Enforcement of China's 'Zero-Covid' Policy," NPC Observer, August 25, 2022; citing Tong Zhiwe, "对上海新冠防疫两措施的法律意见" [Legal Opinion on Two Measures to Prevent the Coronavirus Epidemic in Shanghai], CND.org.

159 **Giorgio Agamben wrote in 2020:** Giorgio Agamben, *Where Are We Now? The Epidemic as Politics* (Rowman & Littlefield, 2021), 8.

161 **locking them inside trembling buildings:** Eva Dou, "Earthquake in China's Sichuan Leads to Outcry over Covid Lockdown," *Washington Post*, September 6, 2022.

161 **Xi grew obsessed:** Dake Kang, "Ignoring Experts, China's Sudden Zero-COVID Exit Cost Lives," Associated Press, March 24, 2023.

162 **The statement carried a sting:** "中共中央政治局常务委员会召开会议 习近平主持会议" [Xi Jinping presides over a meeting of the Standing Committee of the Politburo], Xinhua, May 5, 2022.

162 **broadcast only approved messages:** "禁止转发未经官方证实的和负能量的内容" [Do not forward content that is unconfirmed by official sources

or content with negative energy], *China Digital Times*, September 9, 2022.

162 **One man in Chongqing went viral:** Chris Buckley, Alexandra Stevenson, and Keith Bradsher, "From Zero Covid to No Plan: Behind China's Pandemic U-Turn," *New York Times*, December 19, 2022, updated December 21, 2022.

166 **Economists from Nomura estimated:** Tom Hancock, "China's Regular Covid Testing to Cost 1.8% of GDP, Nomura Says," Bloomberg News, May 6, 2022.

166 **Xinhua released a commentary:** "Is China Really Ill-Prepared for Its New Phase of COVID Response?" Xinhua, January 20, 2023.

167 **Through 2024, every governing party in developed democracies:** John Burn-Murdoch, "What the 'Year of Democracy' Taught Us, in 6 Charts," *Financial Times*, December 30, 2024.

168 **scholarly estimates come to:** Hong Xiao et al., "Excess All-Cause Mortality in China after Ending the Zero COVID Policy," *JAMA Network Open* 6, no. 8 (2023).

Chapter 6: Fortress China

175 **Shanghai's foreign population was in decline:** Jiayao Liu, Gao Yuan, and Zichen Wang, "Sharp Decline in the Number of Foreigners in China Demands Serious Attention," Pekingnology, June 11, 2023.

176 **14,000 millionaires emigrated from China:** James Kynge, "China's Super-Rich Are Eyeing the Exit," *Financial Times*, June 21, 2024.

176 **seen a surge in new homebuyers:** Liam Dillon and Cindy Chang, "This Orange County City Has the Hottest Housing Market in the Country," *Los Angeles Times*, August 16, 2024.

176 **a doubling in the number of Chinese migrants:** Canada had 2,065 investor permanent residency visas in 2019 and 4,020 in 2023. Immigration, Refugees and Citizenship Canada (IRCC), "Canada: Permanent Residents by Country of Citizenship and Immigration Category," Open Government Portal, last updated May 18, 2024. The United States had 3,984 investor visas in FY2019 and 7,464 in FY2024. IIUSA (Invest in the USA), "EB-5 Visa Data Dashboard," last updated December 26, 2024.

179 **Xi Jinping asked the group:** Du Shangze and Li Jianguang, "微观察·习近平总书记在企业和专家座谈会上 '看准了就坚定不移抓'" [Micro observation: "Once certain, firmly pursue it," General Secretary Xi Jinping at an Enterprise and Expert Symposium], *Seeking Truth, People's Daily*, May 28, 2024.

179　**Before their comments were censored:** "【网络民议】主因是什么? 主因就是您啊主席!" [Online public opinion: What's the main reason? The main reason is you, chairman!], *China Digital Times*, May 29, 2024.

180　**Lu subsequently wrote a letter so self-abasing:** Li Tianji, "鲁炜忏悔书在这场展览曝光:妻子对我完全绝望" [Lu Wei's letter of repentance exposed at this exhibition: My wife was completely desperate about me], *Sina News, Beijing Youth*, November 15, 2018.

180　**A former ByteDance executive has publicly accused:** Thomas Fuller and Sapna Maheshwari, "Ex-ByteDance Executive Accuses Company of 'Lawlessness,'" *New York Times*, May 12, 2023.

181　**Zhang Yiming, then CEO, wrote:** David Bandurski, "Tech Shame in the 'New Era,'" China Media Project, April 11, 2018.

181　**Even the Ministry of Education took part:** "China Bans For-Profit School Tutoring in Sweeping Overhaul," Bloomberg News, July 24, 2021, updated July 25, 2021.

181　**lost 90 percent of its market cap:** Sophie Yu and Brenda Goh, "New Oriental Laid Off 60,000 Staff after China's Education Crackdown, Founder Says," Reuters, January 10, 2022.

182　**as Didi wrote after it received:** Paul Mozur and John Liu, "China Fines Didi $1.2 Billion as Tech Sector Pressure Persists," *New York Times*, July 21, 2022.

184　**he unveiled a new leadership team:** Wu Guoguang, "Aerospace Engineers to Communist Party Leaders: The Rise of Military-Industrial Technocrats at China's 20th Party Congress," Asia Society Policy Institute, Washington, DC, February 8, 2023.

184　**In 2018, Xi praised teachers:** Ma Chi, "Xi Jinping: A Model in Respecting Teachers," *China Daily*, September 10, 2021.

185　**"grab little ones from the cradle":** Zhang Changjiang, "革命传统教育要从娃娃抓起" [Education in revolutionary traditions must start from the cradle], *Seeking Truth, People's Daily*, June 25, 2021.

185　**"enter the mind":** "学纪, 知纪, 明纪, 守纪" [Learn discipline, know discipline, understand discipline, obey discipline], *Seeking Truth*, September 2024.

185　**"zero-distance service":** "Setting the Highest Standards, Blossoming the 'Fengqiao Experience' in Beijing: A Summary of the Beijing Public Security Bureau's Work in Creating 'Fengqiao-Style Police Stations,'" Beijing Public Security Department, November 27, 2023.

185　**"We must be prepared for worst-case":** "Xi Urges Accelerated Efforts to Modernize National Security System, Capacity," Xinhua, May 30, 2023.

186　**over business disputes or drug charges:** Jack Wroldsen and Chris Carr, "The Rise of Exit Bans and Hostage Taking in China," *MIT Sloan Management Review*, November 15, 2023.

186 **ten times that many:** John Ruwitch, "Why the Number of American Students Choosing to Study in China Remains Low," NPR, June 13, 2024.

188 **During the height of the pandemic:** "Xi Seeks 'Lovable' Image for China in Sign of Diplomatic Rethink," Bloomberg News, June 1, 2021, updated June 2, 2021.

190 **The comedian, Li Haoshi:** Fan Wang, "China Fines Comedy Troupe $2M for Joke about the Military," BBC, May 17, 2023.

190 **found their shows canceled:** Lyric Li and Vic Chiang, "No Laughing Matter: China Cancels Comedy, Citing 'Force Majeure,'" *Washington Post*, May 25, 2023.

191 **They have gone on a spending spree:** Lingling Wei, "China Reins In Its Belt and Road Program, $1 Trillion Later," *Wall Street Journal*, September 26, 2022.

191 **According to Deloitte:** Deloitte, "Africa Construction Trends 2021," April 5, 2021.

192 **Chinese builders went a billion dollars over:** Neta Cynara Anggina, "Indonesia: The High Cost of High Speed Rail," *Interpreter*, Lowy Institute, November 30, 2023.

192 **Two photographs have circulated:** Eric Olander, "The Number of Leaders Attending China's Belt and Road Forums Has Fallen Steadily," China Global South Society, October 18, 2023.

192 **the World Bank found in 2024:** Lukas Franz et al., "The Financial Returns on China's Belt and Road," World Bank, Annual Bank Conference on Development Economics, July 9, 2024.

192 **developing countries hold China in more positive regard:** Laura Silver, Christine Huang, and Laura Clancy, "China's Approach to Foreign Policy Gets Largely Negative Reviews in 24-Country Survey," Pew Research Center, July 27, 2023.

193 **a more balanced trade relationship:** Jason Douglas, Jon Emont, and Samantha Pearson, "China's Flood of Cheap Goods Is Angering Its Allies Too," *Wall Street Journal*, December 3, 2024.

193 **twenty times more coal-burning capacity:** Molly Lempriere, "China Responsible for 95% of New Coal Power Construction in 2023, Report Says," Carbon Brief, April 11, 2024.

194 **China is demolishing golf courses:** "Xi's Campaign to Feed China Is Turning Wasteland into Farms," Bloomberg News, July 11, 2024.

194 **food insecurity spiked among low-income Americans:** Julia A. Wolfson and Cindy W. Leung, "Food Insecurity and COVID: Disparities in Early Effects for US Adults," *Nutrients* 12, no. 6 (2020): 1648.

196 **few hundred thousand highly trained workers:** Semiconductor Industry Association, "Chipping Away: Assessing and Addressing the Labor Market Gap Facing the US Semiconductor Industry," July 2023.

196 **In 2025, China will graduate:** Remco Zwetsloot, "China Is Fast Out-pacing US STEM PhD Growth," Center for Security and Emerging Technology, Georgetown University, August 2021.

196 **In Xi's telling:** Xi Jingping, "习近平关于科技创新重要论述摘编" [Select important discussions on technological innovation by Xi Jinping], Office of the Leading Group for Studying and Implementing Xi Jinping Thought on Socialism with Chinese Characteristics for a New Era in the Municipal Science and Technology System, April 2023.

196 **"The competition for national strength":** Chinese Communist Party Committee Ministry of Science and Technology Leading Group, "深化科技体制改革 为中式现代化提供强大科技支撑" [Deepen structural reforms in technology, provide strong technological support for Chinese modernization], *Seeking Truth*, September 16, 2024.

198 **Fewer than 1,000 scientists of Chinese descent:** Yu Xie et al., "Caught in the Crossfires: Fears of Chinese-American Scientists," *Proceedings of the National Academy of the Sciences* 120, no. 27 (June 27, 2023).

202 **Yao Yang, a dean at Peking University:** Li Xiaodan, "姚洋: 金融业降薪不是惩罚性的, 而是降低金融业吸引力, 转向发展制造业" [Yao Yang: Cutting salaries in the financial sector isn't punitive, it's to decrease the attractiveness of the financial sector, and develop manufacturing instead], Sina Finances, *Economic Observer*, June 18, 2024.

203 **Meta was about to build a data center:** Hannah Murphy and Cristina Criddle, "Meta's Plan for Nuclear-Powered AI Data Center Thwarted by Rare Bees," *Financial Times*, November 4, 2024.

204 **The quantitative disparities:** John Frittelli, "U.S. Commercial Shipbuilding in a Global Context," Congressional Research Service, November 15, 2023.

204 **In two days, Ukraine could fire as many shells:** Ukraine might fire seven thousand shells. John Ismay, "Pentagon Opens Ammunition Factory to Keep Arms Flowing to Ukraine," *New York Times*, May 29, 2024. America has produced an average of 14,400 shells per month. Roxana Tiron and Billy House, "America's War Machine Can't Make Basic Artillery Fast Enough," Bloomberg News, June 7, 2024.

204 **Jake Sullivan said:** "National Security Advisor Jake Sullivan on Fortifying the U.S. Defense Industrial Base," Center for Strategic and International Studies, December 4, 2024.

206 **Nick Bagley concluded his seminal paper:** Bagley, "Procedure Fetish."

207 **Xi has rejected the idea:** Lucy Hornby, "China's Top Judge Denounces Judicial Independence," *Financial Times*, January 17, 2017.

Chapter 7: Learning to Love Engineers

221 **in 1974, Robert Caro published:** Robert Caro, *The Power Broker: Robert Moses and the Fall of New York* (Knopf Doubleday, 1974).

222 **tore the heart out:** Caro, *Power Broker*, 522.

224 **reckon with Admiral Hyman Rickover:** Marc Wortman, *Admiral Hyman Rickover: Engineer of Power* (Yale University Press, 2022).

225 **embedded in Peter Thiel's quip:** Blake Masters, "CS183: Startup—Peter Thiel Class Notes, Class 11 Notes Essay," Blake Masters (Tumblr), May 11, 2012.

227 **Overall, half of American renters:** "Nearly Half of Renter Households Are Cost-Burdened, Proportions Differ by Race," US Census Bureau Newsroom, September 12, 2024.

227 **It costs five times as much:** America, Second Avenue: $2.5 billion/km (2023 dollars); France, Line One: $457 million/km (2023 dollars). Transit Costs Project, "What the Data Is Telling Us," last updated February 27, 2024.

228 **not a single home has been connected:** "The Harris Broadband Rollout Has Been a Fiasco," *Wall Street Journal*, October 4, 2024.

228 **build electric vehicle charging stations:** Shannon Osaka, "Biden's $7.5 Billion Investment in EV Charging Has Only Produced 7 Stations in Two Years," *Washington Post*, March 29, 2024.

228 **Representative Alexandria Ocasio-Cortez:** William Cummings, "'The World Is Going to End in 12 Years If We Don't Address Climate Change,' Ocasio-Cortez Says," *USA Today*, January 22, 2019.

229 **the classic line by professor Grant Gilmore:** Grant Gilmore, *The Ages of American Law* (Yale University Press, 2014), 99.

231 **Communist Party propaganda organs blared:** Xi Jinping, "习近平: 中国永远是发展中国家的一员" [Xi Jinping: China will always be a member of developing countries], *People's Daily*, August 24, 2023.

232 **ninety-four cars a year:** Odd Arne Westad and Chen Jian, *The Great Transformation: China's Road from Revolution to Reform* (Yale University Press, 2024), 226.

SUGGESTIONS FOR FURTHER READING

America's transition into the lawyerly society (which is not necessarily how the author would describe it) is told in Paul Sabin's *Public Citizens: The Attack on Big Government and the Remaking of American Liberalism* (W. W. Norton, 2021).

An overview of American legal culture is Robert Kagan's *Adversarial Legalism: The American Way of Law* (Harvard University Press, 2003).

The scale of China's construction and how its political economy works is discussed in Arthur Kroeber's *China's Economy: What Everyone Needs to Know* (Oxford University Press, 2nd edition, 2020).

For how China fits into Asia's growth story, there is no book more compelling than Joe Studwell's *How Asia Works* (Grove Press, 2014).

The essential views of the Industrial Party are laid out in Liu Cixin's *Three-Body Problem* trilogy (Tor Books, 2016).

The story of Song Jian's influence is told in Susan Greenhalgh's *Just One Child: Science and Policy in Deng's China* (University of California Press, 2008).

Child abandonment and its heartbreaking consequences are laid out in Kay Ann Johnson's *China's Hidden Children: Abandonment, Adoption, and the Human Costs of the One-Child Policy* (University of Chicago Press, 2017).

The most interesting account of life in China during enforcement of zero-Covid is Peter Hessler's *Other Rivers: A Chinese Education* (Penguin, 2024).

The most vital book about escape from Han rule, spanning thousands of years, is James C. Scott's *The Art of Not Being Governed: An Anarchist History of Upland Southeast Asia* (Yale University Press, 2010).

Why should America build? See Ezra Klein and Derek Thompson's *Abundance* (Avid Reader Press, 2025).

Perhaps the best China book of all is *Fuchsia Dunlop's Invitation to a Banquet: The Story of Chinese Food* (W. W. Norton, 2023).